TABLE OF CONTENTS

Cosmogenic nuclides — 45

Anthropogenic radionuclides — 65

Materials Sciences

Education and outreach

EDITORIAL

Research at the Laboratory of Ion Beam Physics (LIP) continued to demonstrate an outstanding track record in 2022. Thus, it is once again a great pleasure for us to present this annual report, which is published in its present form for the 14th time. In the past year, the breadth of research fields we worked on was impressive and shows that the scope of our research activities extends well beyond the fundamental physical sciences.

With the instrumental developments of our AMS spectrometers, we have reached a level of precision that allows a whole new quality of radiocarbon dating measurements that, at the discovery of the method, people did not dare to dream of. The annually resolved calibration data now available in high quality from tree ring chronologies have contributed significantly to this. It is also particularly remarkable that with our latest low energy AMS system LEA we have succeeded in performing measurements of highest precision in routine operation. Dedicated to CO_2 gas analyses, the prototype MICADAS system serves as the backbone of compound specific radiocarbon analyses at LIP. The Tandy instrument, the mother of all compact AMS systems relying on collisional molecular breakup reactions, is now exclusively in use for ^{129}I measurements. The MILEA multi-isotope AMS system is serving the needs for ^{10}Be, ^{26}Al and actinide analyses. Superb ^{36}Cl and experimental ^{32}Si measurements highlight the unabated importance of the 6 MV Tandem. In addition, a very successful program for material analyses with various ion beam methods is running on our 1.7 MV Tandetron. Thus, we now have a whole instrument park at our disposal which ensures flexible operation, now and in the future.

A special highlight of the past year was the 24th Radiocarbon Conference, which we hosted together with the 10th Symposium on Radiocarbon and Archaeology here at ETH. With more than 400 registered participants, the event was a huge success. It was a great pleasure to welcome researchers from over 35 nations who were enthusiastic about scientific exchange after years of travel restriction.

Once again, we would like to thank all our partners, collaborators, and our scientific, technical, and administrative personnel. Each and every one of you makes a significant contribution to the success of our laboratory and your dedication is an important pillar for the smooth operation of LIP.

Hans-Arno Synal and Marcus Christl

THE LIP ACCELERATOR FACILITIES

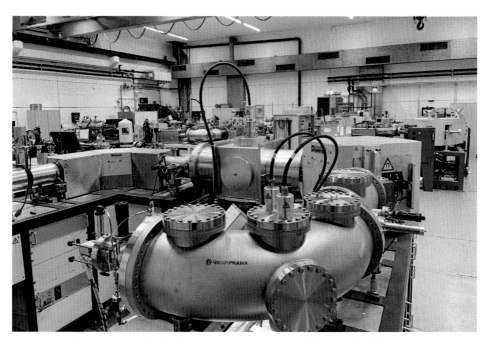

View into the measurement hall at various instruments (front to back): LIP-MICADAS, Microbeam Beamline of Tandetron, MILEA, Tandy, Proto-MICADAS and gas-filled magnet of the 6 MV Tandem

Activities on the 6 MV Tandem system

Tandetron operation

Activities on the 0.5 MV Tandy system

Activities on the 0.3 MV MILEA System

Routine ^{10}Be measurements on MILEA

^{14}C on the 200 kV MICADAS and the 50 kV LEA

ACTIVITIES ON THE 6 MV TANDEM SYSTEM

Beam time and sample statistics

C. Vockenhuber, scientific and technical staff of Laboratory of Ion Beam Physics

In the first half of 2022 the 6 MV Tandem accelerator was running well. The beamtime statistics in Fig. 1 shows a similar picture as in the last years. Most of the 700 beam time hours was used for AMS measurements that require high energy for isobar separation (^{36}Cl and ^{32}Si). About half of it was used for routine measurements of ^{36}Cl with the Gas-filled Magnet; the other half for developments of the measurement setup of ^{32}Si. A small fraction of the beamtime was used for implantation experiments of ^{27}Al into SiC at energies of 30 and 48 MeV in collaboration of Advanced Power Semiconductor Lab at ETH.

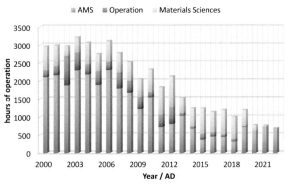

Fig. 1: *Time statistics of the Tandem operation subdivided into AMS (blue), materials sciences and MeV-SIMS (green), and service and maintenance activities (red).*

However, during ^{36}Cl AMS measurements it became apparent that the stripper foils had been used up and need to be replaced. After a problem with the current reading of the LE column occurred in October we finally decided to open the accelerator after more than three years of operation. In the following tank opening we did a major revision of the accelerator including of service of the column resistors. In particular the Pelletron charging system was overhauled in many ways. We replaced several bearings in drive motors, charging wheels and pick-up pulleys. Also, the GVM motor got new bearing

ensuring a long and smooth running in the coming years. Finally, a new batch of more than 40 newly floated 2 µg/cm^2 stripper foils were installed.

Fig. 2: *Decontamination of the gas dryer at refurbish company Suter Zotti AG (left) and the old heating elements with asbestos construction bars (right).*

An additional problem occurred in the gas handling system for the tank gas. To ensure the high insulation strength of the CO_2-N_2 mixture the gas has to be very dry and needs to be dried in an external gas dryer a few times per year. However, the device, still from the original installation in early 1960s, failed in summer of 2022. During the repair insulation material made from asbestos was found requiring a special decontamination of the interior and complete overhaul of the gas dryer (Fig. 2). As of January 2023, the gas dryer is still at local company specialized in building custom-made heating elements. Thus, planned beam times for the end of the year had to be postponed to 2023.

In parallel we prepared most of the components for the upgrade of the LE side of the accelerator with a MICADAS-type ion source and an improved injection layout that is planned for 2023.

TANDETRON OPERATION

Materials science at the 1.7 MV Tandetron

A.M. Müller, C. Vockenhuber, scientific and technical staff of Laboratory of Ion Beam Physics

In 2022 the 1.7 MV Tandetron facility and its beamlines were used for most of our ion beam analysis activities. Although it was operational most of the time in 2022, some maintenance work was necessary and caused some downtime. Nevertheless, we could measure more than 900 samples for 50 different users. With the growing number of users, we started to implement a ticketing system for the communication between users and experimentalists. The system is based on Request Tracker, an open source enterprise grade ticketing system hosted by ISG.

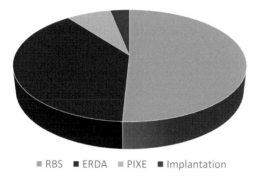

RBS ■ ERDA ■ PIXE ■ Implantation

Fig. 1: Breakdown of the 912 analyses sorted by applied technique.

The beamtime distribution in Fig. 1 shows that about 50% of the samples are analyzed by Rutherford Backscattering Spectrometry (RBS), which are complemented by Particle Induced X-Ray Emission (PIXE) measurements in 7% of the cases. The demand for Elastic Recoil Detection Analysis (ERDA) measurements increased from 27% in 2021 to 40% in 2022, a continuing trend observed over the last years. Implantations for various applications take up only a few %.

Unfortunately, the RF driver of the Tandetron failed again similar to the end of 2021 and needed to be send to HVEE for repair.

For RBS measurements we use a He beam from the duo-plasmatron ion source. The filament for the plasma ignition shows limited lifetime and needs frequent recoating or if it breaks a complete replacement with a new Pt gauze which is costly. Furthermore, the coating process needs a long time for outgassing and delays measurements. Thus, we spent some time to test the source with a filament made out of a Mo wire (Fig. 2). We could run the ion source with a simple Mo wire even without coating. Although replacement is simple and fast, the Mo can only be used once before it breaks. Therefore, we continue to use the coated Pt gauze for the moment. Further tests and simulations with different filament materials and geometries are planned.

Fig. 2: Mo filament for the Duo-plasmatron ion source, unfortunately broken after a few hours of operation.

ACTIVITIES ON THE 0.5 MV TANDY SYSTEM

Beam time and sample statistics

M. Christl, scientific and technical staff of Laboratory of Ion Beam Physics

In 2022, the Tandy system was almost exclusively used for the AMS analysis of ^{129}I in sea water samples (Fig. 1). The group members of Physical Oceanography at D-USYS, ETH Zurich led by Prof. Casacuberta were almost independently running the system for ^{129}I and they were also involved in the revision and cleaning of the accelerator tank in December. Also, the refilling of the He stripper gas bottle became necessary during this revision.

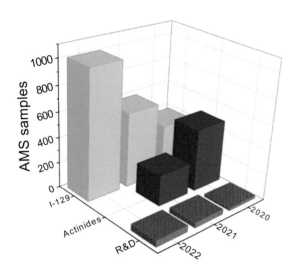

Fig. 1: *Annual number of samples measured on the TANDY over the past three years.*

This year all actinide measurements were transferred to the MILEA system (Fig. 1). As a consequence, ^{129}I analysis make up almost all of the more than 1000 samples analyzed in 2022 (Fig. 2). The analysis of ^{129}I in ocean water are performed to better understand the origin of water masses in the Arctic and North Atlantic Ocean and to assess timescales of water mass circulation. Results of these studies are presented in several reports in the section of anthropogenic nuclides in this volume.

In addition to that, the Tandy was used for R&D studies related with the improvement of the gas

ionization chambers that are used in our AMS systems to detect the rare radionuclides. In this context, different gas types were studied with respect to energy resolution and signal intensity. The results of this study are presented in a separate report in the technical development section of this volume.

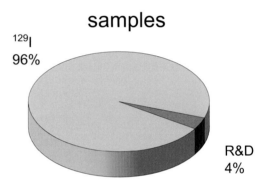

Fig. 2: *Relative distribution of the number of AMS samples measured on the Tandy in 2022.*

Almost four fifth of the total beam time of 560 h was dedicated to ^{129}I analyses (Fig. 3), while the remaining fifth was spent for R&D projects to improve our detection systems.

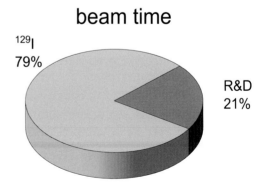

Fig. 3: *Relative distribution of the Tandy beam time in 2022.*

ACTIVITIES ON THE 0.3 MV MILEA SYSTEM

Beam time and sample statistics

M. Christl, P. Gautschi, scientific and technical staff of Laboratory of Ion Beam Physics

In 2022 the MILEA system was in operation for more than 3500 hours. While ^{10}Be, and ^{26}Al are routinely measured on MILEA since a few years, this year all actinide measurements have been performed on the system, and first routine measurements of ^{129}I have been performed. With this, the MILEA system was more and more used as multi isotope facility with most of the beamtime allocated for ^{10}Be and almost equal beam time distributions for ^{26}Al and the actinides (Fig. 1).

Technical University (CTU) within the context of a RADIATE project.

In 2022 MILEA has revealed it full multi isotope capabilities and with steeply rising sample numbers (Fig. 3) it is on a good way to become our new working horse for all isotopes (besides ^{14}C and ^{36}Cl).

AMS samples

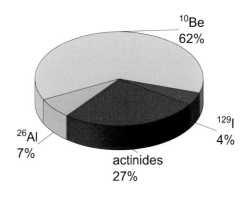

Fig. 2: Relative number of AMS samples measured in 2022.

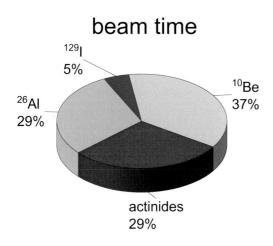

Fig. 1: Relative beam time distribution for the different operations on MILEA in 2022.

In 2022, a total of more than 2100 user samples were analyzed for various nuclides on MILEA (Fig. 2 and Fig. 3). The majority of more than 1300 samples were analyzed for ^{10}Be covering many different projects from ice core and marine sciences to in-situ dating. More than 550 samples were analyzed for U-, Pu-, Am-, and Cm-isotopes covering various projects such as marine tracer studies, nuclear forensics, and source apportionment of contaminated soils. First ^{129}I measurements were carried out in sea water samples in collaboration with the Czech

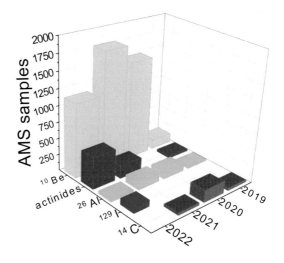

Fig. 3: Temporal evolution of AMS samples (unknowns) measured on MILEA.

ROUTINE ^{10}Be MEASUREMENTS ON MILEA

Currents, sample types and efficiency

M. Christl, P. Gautschi, H.-A. Synal

In 2022, a total of 1333 user samples were analyzed for ^{10}Be on MILEA. The current distribution of all user samples (Fig. 1) shows a large variation which is mainly caused by different amount of Be-carrier added to the samples (100 - 500 µg) and variable yields during chemical sample preparation in the user labs. The median average BeO$^-$ sample current of all user samples was 4.6 µA indicating that the MICADAS type ion source was running fairly well. This resulted in rather short measurement times of 30 min per sample (median) ranging from about 10 min to almost 2 hr per sample.

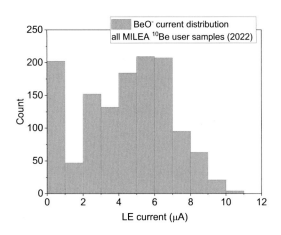

Fig. 1: *Distribution of the average BeO$^-$ currents of all ^{10}Be samples measured on MILEA in 2022.*

This year, the ^{10}Be samples came from nine different laboratories from Europe, China, and India. More than half of all samples (Fig. 2) was analyzed from polar ice cores to detect solar events and to study past solar variability. In about 20% of the samples in-situ produced ^{10}Be was analyzed in order to date moraines, to quantify paleo-erosion rates of landscapes, or to determine burial ages of deposits. Another 20% of the samples came from marine sediments where ^{10}Be is used as a tool for synchronization with other climate records. The remaining 6% of

analyses were performed in precipitation (rain/snow) samples, or to determine basin wide erosion rates with meteoric ^{10}Be.

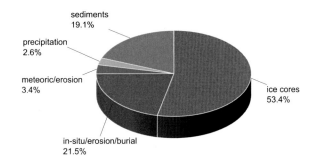

Fig. 2: *Types of ^{10}Be samples analysed in 2022.*

The total number of ^{10}Be atoms in the samples varied by more than four orders of magnitude from several 10^4 to $>10^9$ (Fig. 3). The final uncertainty is dominated by counting statistics for samples containing less than 10^7 at of ^{10}Be. We estimate that our over-all efficiency may reach values of slightly more than 1‰ (Fig. 3).

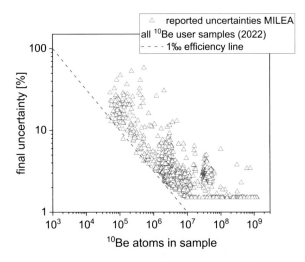

Fig. 3: *Final uncertainty of all reported ^{10}Be data in 2022 vs. ^{10}Be atoms in the sample. The red dashed line indicates a detection efficiency of 1‰.*

^{14}C ON THE 200 KV MICADAS AND THE 50 KV LEA

Performance and sample statistics

L. Wacker, N. Haghipour, U. Ramsperger, scientific and technical staff of Laboratory of Ion Beam Physics

The radiocarbon samples were measured for the first time on 3 instruments in 2022. In February 2022 the new 50 kV LEA system went into operation and already 1 1/2 months later in mid-March first routine ^{14}C measurements were performed [1]. Consequently, about 1/4 of all graphite samples were measured on the LEA system (Fig. 2). Primarily a shortage in laboratory technicians was responsible for an almost 10% lower quantity of graphite samples being measured, although the measurement capacities were increased. This meant primarily less samples were measured for our internal projects. Nevertheless, more than 7000 graphite targets were measured, which is the second highest number ever.

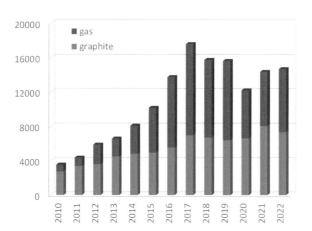

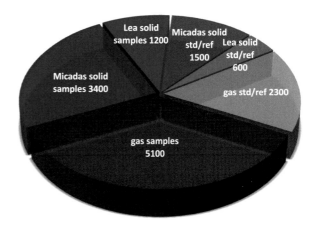

Fig. 2: *Overview of all samples measured on the two MICADAS and the new LEA system in 2022 at LIP. Graphite samples measured on the LIP-Micadas and the LEA are given in shades of blue. Most samples were measured as gas on the Proto-Micadas only (red/orange).*

Fig. 1: *The number of total measured radiocarbon samples in 2022 was slightly higher than in 2021. While the graphite samples measured declined by nearly 10%, more than 15% more gas samples were measured.*

The measured gas samples on the Proto-Micadas over-compensate the reduction of graphite samples with an increase of more than 1000 targets. This is a very high number compared to previous years. Only in years, when biomedical samples running at a higher pace were analyzed, more gas samples were measured. Overall,

14,100 measurements were performed, of which 9,700 were unknowns and 4,400 were standards, blanks or references.

The normally relatively constant fraction of primarily gas samples measured for our partners in the Earth Sciences Department, D-ERDW, increased in 2022 unexpectedly about 20% to an all-time high of 3200 samples. In addition, also 10% more gas samples (1000) were measured on the base of research contracts.

[1]   U. Ramsperger et al., LIP Annual Report (2022) 16

14

INSTRUMENTAL AND ANALYTICAL DEVELOPMENTS

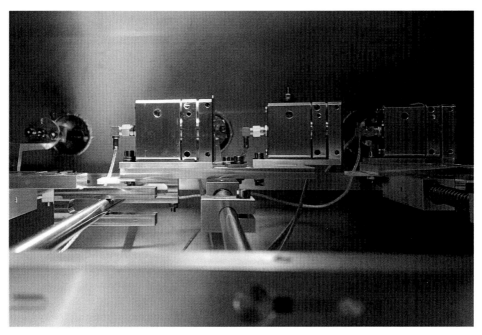

Multi-Faraday Cup arrangement at the HE side of MILEA

LEA - a low energy accelerator for ^{14}C dating

A ^{14}C detector for LEA

Online Ramped Oxidation (ORO)

Low-level ^{129}I at MILEA

Efficiency determination of ^{32}Si detector setup

The art of floating thin stripper foils

LEA - A LOW ENERGY ACCELERATOR FOR ^{14}C DATING

Stability test performance in comparison with MICADAS

U. Ramsperger, D. De Maria, P. Gautschi, S. Maxeiner[1], A.M. Müller, H.-A. Synal, L. Wacker

A newly developed Accelerator Molecular Spectroscopy (AMS) system, LEA (Low Energy Accelerator, Fig. 1), is tested and compared with the high-quality AMS system MICADAS (Mini CArbon DAting System). The main difference between these two systems is the acceleration voltage, which is reduced from 200 kV with the MICADAS system to 50 kV with the LEA system.

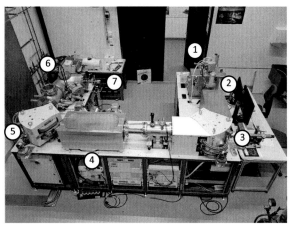

Fig. 1: *LEA system consisting of the following components: sample changer (1), ion source (2), low energy magnet (3), accelerator (4), high energy magnet (5), electrostatic deflector (6), ^{14}C detector (7).*

LEA vs MICADAS: As a performance test, the LEA system is compared with the MICADAS system, a well-established and highly accurate system with a precision and stability performance of less than 1‰ [1]. To carry out the tests two sets of samples, each composing of 7 standards, 4 blanks and 26 wood samples, are measured on both systems. The wood samples are consecutive annual rings of a tree of known age. The experiments are taken as follows: First, one set of samples are measured on the LEA system shown in Fig. 2 (cal BP 2800 – 2825, Data LEA 1). Then the very same set of samples are transferred to the MICADAS system and are measured equally (Data MICADAS 1), and vice

versa for the second set of samples (cal BP 2826 – 2851, Data LEA 2 & MICADAS 2). In other words, the second set of samples are measured on MICADAS before they are measured on LEA. The agreement of the results is striking, as shown in Fig. 2, with a mean deviation of these two AMS measurements of only 1.0 ± 4.7 year. Please note that in these measurements, the first runs of each sample after insertion into the respective system are omitted as the sample materials have been exposed to air during the transfer of the magazines and therefore their surface may have become contaminated.

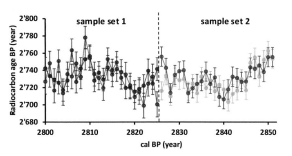

Fig. 2: *Dark blue (Data MICADAS set 1) and bright blue (Data MICADAS set 2) dots and lines in the diagram indicate data points of MICADAS measurements, while dark red (Data LEA set 1) and bright red (Data LEA set 2) dots and lines show data points of LEA measurements. The horizontal axis represents the (known) calendar age of the samples and the vertical axis represents the measured radiocarbon age. The results of these measurements on these two independent AMS systems are in excellent agreement.*

[1]   L. Wacker et al., NIM-B 455 (2019) 178

[1]*Ionplus AG, Dietikon*

A ^{14}C DETECTOR FOR LEA

A Bragg type gas ionization detector operated in the proportional region

U. Ramsperger, D. De Maria, P. Gautschi, S. Maxeiner[1], A.M. Müller, H.-A. Synal, L. Wacker

A Bragg type gas ionization detector run with isobutane 3.5 is used to detect the ^{14}C ions. Since the energy of the ^{14}C ions is significantly lower than in the MICADAS system (less than 150 keV), the detector [1] is operated in the proportional counting region instead of the ion chamber region. Several test measurements taken with the LEA system show that values of 450 V for the applied voltage and a gas pressure of 9.5 mbar provide the most accurate results in separating the background from the real ^{14}C events. Since the ^{14}C isotope energy is only 140 keV the detector signal has to be amplified by a factor of approximately three, resulting in a signal level of 3 V. By comparison, the MICADAS uses 300 V detector voltage at an isobutane gas pressure of 20 mbar. With the same settings of the amplifier electronics and total ^{14}C isotope energy of 420 keV and an amplification factor of one, the signal level is equally 3 V. The thickness of the entrance window separating the vacuum and the isobutane gas of the detector also plays an important role in ensuring a clean division between the background and the ^{14}C events. Two different Si_3N_4 detector windows are tested with nominal thicknesses of $d_1 = 30$ nm and $d_2 = 50$ nm, respectively. Test measurements were conducted using graphitized Phthalic acid (Pha) blank samples as target materials. They are presumably free of ^{14}C due to their petrochemical origin but usually contain a ^{14}C contamination record from the preparation process which is typically visible at a low 10^{-15} ratio with respect to ^{12}C. Therefore, the obtained spectra show a distinct ^{14}C peak and

background events which predominantly dissipate lower energy signals in the active detector volume. The background events and the ^{14}C events in the energy spectrum of the 50 nm window, indicated as a red line in Fig. 1, converge and cannot be properly separated. Conversely, due to the lower stopping energy and energy straggling [2] in the thinner 30 nm membrane window, the energy spectrum recorded, marked with a blue line in Fig. 1, shows two well-separated peaks that clearly distinguish the ^{14}C events from other background events.

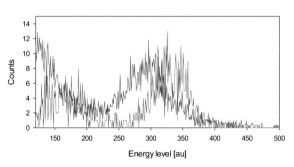

Fig. 1: *Two energy spectra of a Pha blank sample taken with an isobutane gas detector through 30 nm (blue lines) and 50 nm (red lines) Si_3N_4 membrane windows respectively. Each spectrum is measured with the same blank sample for 30 min at a high energy ^{12}C current of >10 μA. The gap between the background peak and the ^{14}C peak is clearly visible in the measurement taken with the 30 nm window, while the two peaks overlap in the spectrum with the 50 nm window.*

[1] A. M. Müller et al., NIM-B 356-357 (2015) 81

[2] G. Sun et al., NIM-B 256(2) (2007) 586

[1]Ionplus AG, Dietikon

ONLINE RAMPED OXIDATION (ORO)

A novel set up for coupled thermal and ^{14}C dissection of organic matter

M. Bolandini[1], L. Bröder[1], D. De Maria, T.I. Eglinton[1], L. Wacker

The combination of serial or ramped pyrolysis and/or oxidation to radiocarbon isotope measurements provides insights into the relationships between thermal activation energies and age distributions of natural organic materials [2]. To establish this technique at the Laboratory of Ion Beam Physics, a newly designed system for online ramped oxidation (ORO) has been developed. It is coupled to a double zeolite trap interface (DTI, [1]) to sequentially collect CO_2 from ORO and afterward release it to the gas-accepting ion source of a MICADAS AMS system (schematic in Fig. 1).

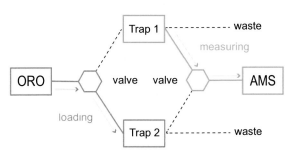

Fig. 1: *Schematic representation of the ORO-DTI-AMS coupling for online sequential combustion and ^{14}C measurement of natural samples. The CO_2 released by combustion (ORO system) is collected on zeolite traps (DTI) and ultimately analysed for [14]C (AMS).*

This innovative method allows to disentangle organic constituents according to their thermal lability by measuring small CO_2 fractions for ^{14}C from combusted sediments at a constant temperature ramp. The combustion of the bulk sample is achieved via a two-stage oven (a ramping oven with constant temperature rate and a combustion oven at fix temperature of ca. 850°C) that regulates the decomposition and the later oxidation of carbonaceous compounds to CO_2 in a mixture of 10% O_2 in He. The coupling to the DTI enables a continuous sampling thanks to the alternating use of its two external traps,

allowing for parallel loading and measuring. Prior to the actual ^{14}C measurements, a pre-screening run is performed to select the suitable trapping intervals for fraction collection. See an exemplary thermogram for Swiss standard soil (combustion at a temperature ramp of 9°C per min) in Fig. 2. The carbon fractions corresponding to specific temperature intervals (black dotted lines in Fig. 2) are defined so that the CO_2 yield suffices for ^{14}C analysis (aim ≥ 10 μg C).

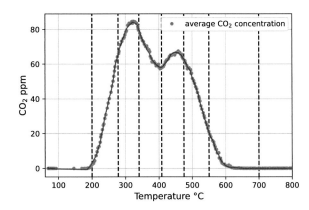

Fig. 2: *Thermogram for in-house reference material Swiss standard soil with a combustion temperature ramp of 9°C per min.*

The performance of the ORO system will be evaluated in a series of tests addressing the effect of sample amount and ramping speed on both thermogram shape and ^{14}C measurements. Successively, standards, as well as reference sample materials will be analysed for comparison with established systems like the RPO from NOSAMS [2].

[1]   D. De Maria et al., Drug Metab. Pharmacokinet. 39 (2021) 100400

[2]   J. D. Hemingway et al., Radiocarbon 59 (2017) 179

[1]Earth Sciences, ETH Zurich

LOW-LEVEL ^{129}I AT MILEA

Determination of ^{129}I background with a clean ion source

C. Vockenhuber, M. Mindová[1]

Our MILEA (Multi Isotope Low Energy AMS) has not been used for ^{129}I measurement so far apart from tests during the initial commissioning phase. This situation allowed us now to assess the background for ^{129}I of the ion source which hasn't seen much ^{129}I. With minimal tuning with the low standard E1 we started measuring low-level Woodward Iodine (WWI) samples. A ^{129}I/^{127}I ratio of $(18 \pm 2) \times 10^{-15}$ (Fig. 1) was measured which is lower than measured at the Tandy $(34 \pm 3) \times 10^{-15}$ [1] and close to the lowest reported value of Woodward material $(13 \pm 1) \times 10^{-15}$ [2].

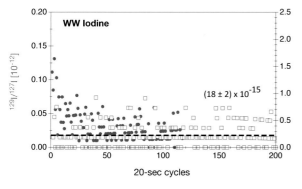

Fig. 1: *Initial measurement of a WWI sample (filled circles) showing cross-talk at the beginning from the tuning sample; a subsequent sample (open squares) gives very low ratios.*

We performed scans of critical devices to investigate possible background. Compared to the Tandy [1] the intensity of ^{127}I at the HE side with same p/q or E/q as ^{129}I is reduced. However, the spectrometers of both instruments are good enough to remove these ions.

Stripper scans (Fig. 2) reveal that molecules can contribute to the background. At low stripper pressures WWI shows molecular background that is interpreted as ^{127}IH$_2$. On the contrary samples mixed with Nb pressed into Cu cathodes have a much stronger background which requires higher stripper pressure to destroy sufficiently (Fig. 2). These molecules do not scale with the ^{127}I

current and thus are not monitored during a measurement. So, care must be taken to choose the correct stripper pressure for samples with such a molecular background.

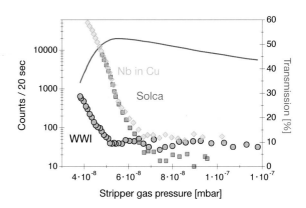

Fig. 2: *Stripper scans showing transmission (red) and count rates of WWI (grey circles), Solca (blue squares) and Nb in Cu (green diamonds).*

Having molecular background in control the remaining contribution is cross-talk in the ion source from samples with high ^{129}I content requiring to keep the ^{129}I/^{127}I ratios within a factor of 100. The very low ratios measured at the beginning of this work were later never achieved again due to long-time cross-talk from higher samples [3].

^{129}I can be measured at MILEA with a similar performance as on the Tandy, the transmission is around 50% for the 2+ charge state, currents are generally limited to less than 10 µA. The random access of samples and the possibility to remove high samples from the magazine without turning the ion source off is a big advantage of MILEA.

[1] C. Vockenhuber et al., NIM-B 361 (2015) 445

[2] L.K. Fifield et al., NIM-B, 530 (2022) 8

[3] M. Mindová et al., LIP annual report (2022) 77

[1]*Nuclear Chemistry, CTU Prague, Czech Republic*

EFFICIENCY DETERMINATION OF ^{32}SI DETECTOR SETUP

Towards the absolute measurement of ^{32}Si

M. Schlomberg, C. Vockenhuber, H.-A. Synal

Within the SINCHRON collaboration, it is planned to redetermine the half-life of ^{32}Si. LIP will perform the measurement of the atom density of ^{32}Si in an aliquot of the master solution of which the specific activity is measured by decay counting. To perform this absolute AMS measurement, it is essential to separate ^{32}Si from its isobar ^{32}S which is achieved at 30 MeV using ^{32}Si in the 4+ charge state from the Tandem facility. A passive absorber gas cell is installed in front of a gas ionization detector, stopping the ^{32}S beam and allowing to detect ^{32}Si. An optimization of the detector setup was performed to minimize background originating from light recoils produced in the entrance foils of the absorber cell and the detector [1, 2, 3]. However, to measure absolutely, it is crucial to know the efficiency of the detector setup.

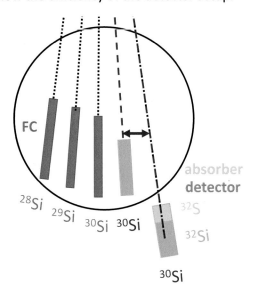

Fig. 1: *Setup to measure the efficiency of the ^{32}Si detector setup.*

To tackle this question, a second detector is installed inside the chamber behind the high energy magnet (Fig. 1). This setup enables to send the same predefined beam of attenuated ^{30}Si in both detectors. The detector without an absorber in front is installed inside the chamber and located ion-optically at the same position as the Faraday cups allowing to measure the complete beam intensity. Consequently, this count rate can be compared to the one measured with the ^{32}Si detector setup with the absorber in front. Fig. 2 shows the measured efficiency in dependence to the applied absorber cell pressure. The beam widens up due to scattering in the stopping process inside of the absorber cell resulting in parts of the beam being cut off in the detector entrance foil and therefore, reducing the efficiency.

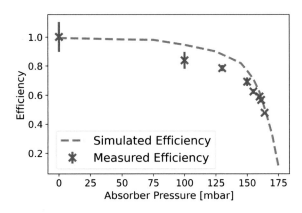

Fig. 2: Measured efficiency and simulated with SRIM vs. applied absorber cell pressure.

The beam losses in the absorber-detector setup are simulated with SRIM. The measured data are in good agreement to the simulated efficiency. In the current setup, an Ar pressure of 154 mbar is needed to separate ^{32}Si from ^{32}S which lies in the rather steep flank of the efficiency curve indicating that an improvement or exact quantization of the efficiency is needed.

[1] M. Schlomberg et al., LIP Annual report (2021) 21

[1] M. Schlomberg et al., LIP Annual report (2021) 22

[2] M. Schlomberg et al., NIM-B 533 (2022) 56

THE ART OF FLOATING THIN STRIPPER FOILS

Improving the transmission for [36]Cl measurements at the 6 MV Tandem

C. Vockenhuber

The transmission through the tandem accelerator is one of the critical parameters for the AMS performance. In contrast to low energy AMS, where gas strippers are commonly used, foil stripping results in an increased charge state and thus in a higher achievable energy with high transmission. However, the drawback of using stripper foils is the limited lifetime and degradation of the measurement time.

For [36]Cl AMS measurements we need high energy for the separation of [36]Cl from its isobar [36]S and thus we use the charge state 7+ at 6 MV which is close to the maximal yield for this charge state. Typically, very thin carbon foils with a thickness of a few $\mu g/cm^2$ are used. We have made very good experience with laser-ablation (LPA) foils obtained from the Technical University in Munich. These foils come on glass slides with a betaine and copper layer in the between and have to be floated onto the stripper foil holders in a multistep process (Fig. 1): cutting foils on the slides, slowly floating on the surface of water, transferred to a HNO_3 bath to remove the protective Cu layer and finally transferring it to water bath from where the foils are fished with the sample holders. The success rate can be very low, but with some experience we managed to achieve a rate of >90% for some slides.

Fig. 1: *The delicate process of floating stripper foils (transfer from the first water to the HNO₃ bath to remove the protective Cu layer).*

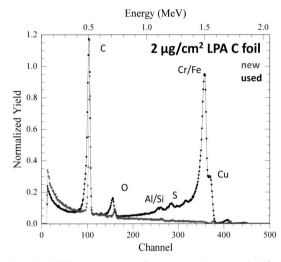

Fig. 2: *RBS spectrum of a new and a used C-foil.*

Several years ago we switched from 3 $\mu g/cm^2$ to 2 $\mu g/cm^2$ foils resulting in an improved transmissions of >20% for Cl^{7+}. However, the transmission steadily drops by ~30% over a measurement day, resulting in several stripper foil changes during a measurement week. In addition also the energy loss increases, which is compensated automatically by increasing the terminal voltage so that the ions have again the same magnetic rigidity at the high energy side. This observation suggests that the foil thickness increases under beam irradiation rather than holes in the foils are produced. With increased thickness angular scattering increases resulting in loss in transmission (and also further down the beamline in the Gas-Filled magnet).

RBS analysis of new and used foils (Fig. 2) clearly show that the C foils not only increase their thickness but also that a significant amount of heavier elements (up to 10%) are deposited onto the foil. Mainly elements around Fe are deposited indicating that the beam might kick out atoms from stainless steel beamline components (in particular the stripper canal in front of the stripper foils), but its origin is still unclear.

RADIOCARBON

Separation of root and macro remains

Organic carbon in the deep Sargasso Sea

Coupled ^{14}C and ^{13}C measurements of SPA-127

Radiocarbon analysis of fossil pollen

^{14}C PREPARATION LAB IN 2022

Overview of samples and laboratory activities

I. Hajdas, A. Albrecht, M. Alter, N. Brehm, D. De Maria, P. Gautschi, Y. Gu, G. Guidobaldi, N. Haghipour, K. Kündig, G. Scacco, H.-A. Synal, L. Wacker, M. Wertnik, C. Welte, K. Wyss

Our laboratory is involved in numerous interdisciplinary research projects and provides analysis for external users (Fig. 1). Dependent on the type of material and its condition, samples submitted to for ^{14}C analysis require different preparation procedures. The preparation step might require only graphitization such as is the case of CO_2 extracted from ground water and submitted in glass tubes. The preparation of water samples to extract the CO_2 from dissolved inorganic carbon DIC [1]. Nearly 400 ocean water samples (Tab. 1) were prepared as a part of collaboration with D-USYS.

Research	Targets
Archaeology	1028
Art	296
Bomb peak	63
Calibration	1032
Climate	215
Environment	48
Geochronology	1523
Ocean	1090
Water	391
Other	151
Reference material	311
Total	**6148**

***Tab. 1:** Number of targets analysed after preparation by the LIP ^{14}C laboratory in 2022.*

Geochronology and archaeology are the leading applications with the most of samples submitted to the LIP laboratory. Some samples require more than one analysis, when for example a plant remains or pieces of charcoal are selected after sieving of sediments or peat. For such samples ^{14}C analysis are performed on total organic carbon TOC of fine fraction as well as on the identified organic fragments. Often, the resulting radiocarbon ages differ but combined information is helpful in interpretation.

The last year was marked by a come back to a normal laboratory modus, which allowed work of students from the ETH and other Swiss universities. Visitors from China, Poland and Czech Republic were able to complete their specific projects dedicated to dating sedimentary records and mortars.

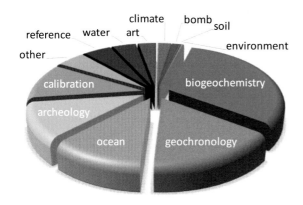

***Fig. 1:** ^{14}C analysis for various research disciplines, during the last year.*

Other type of samples needs totally different approach. Often treatment and purification of sample is needed before to the final separation of CO_2 by combustion or dissolution in acid. Preparation involves visual investigation, sieving, selection of macro remains, removal of visible contamination such as roots. In our laboratory the invisible contaminants are detected using infrared spectroscopy (FTIR). Pre-screening and controlling of sample purity after the treatment is a crucial factor in measuring accurate ^{14}C concentration of the sample.

[1]   N. Casacuberta et al., Radiocarbon 62 (2020) 13

SOLAR ACTIVITY RECONSTRUCTION FROM RADIOCARBON

Analysis of two ancient solar minima by using radiocarbon in tree rings

N. Brehm, M. Christl, L. Wacker, A. Bayliss[1], K. Nicolussi[2]

Solar variability affects the Earth's climate on long (multi-millennial) and short (decadal) timescales. It is, however, not straightforward to investigate the sun-climate connection directly since the magnitude of solar irradiation changes is small and the period of high precision satellite-based measurements covers not more than 50 years. Beyond that period, documented sunspot numbers reaching back to 1610 AD are used as an indicator for solar activity. Going back even further in time the reconstruction of solar activity has to rely on proxy data such as cosmogenic radionuclides [1].

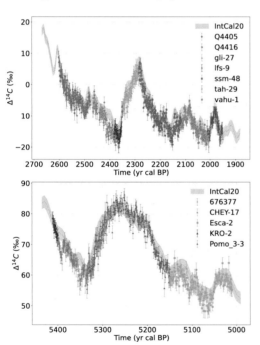

Atmospheric concentration of ^{14}C is recorded in tree rings and can be measured via Accelerator Mass Spectrometry (AMS) over the last 10'000 years. This information can be used to gather information about past solar shielding from cosmic rays of the sun. Here, two annually resolved and accurately dated records of atmospheric ^{14}C from trees are used to provide temporally accurate reconstructions of ^{14}C production that enables the systematic investigation of solar variability over two time regions around 2600 BP to 1950 BP and 5000 BP to 5600 BP (Fig. 1).

The two time periods were analysed with a global carbon cycle box model to reconstruct past ^{14}C production rate, which is needed to get the solar modulation parameter to describe past solar activity. Both newly analysed time periods cover a prolonged phase of reduced solar activity. The new data allows for a direct comparison of these minima with the more recent Spörer minimum (Fig. 2). First analysis shows a similar but not as long drop in solar activity during the newly analysed solar minima.

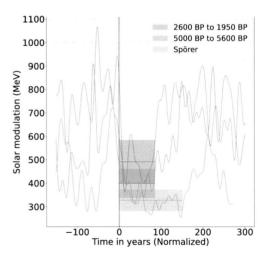

Fig. 1: Annual ^{14}C data from tree rings covering the time periods around 2600 BP to 1950 BP and 5000 BP to 5600 BP

Fig. 2: Solar modulation parameter from three different annually resolved ^{14}C records.

[1] R. Muscheler et al., Quat. Sci. Rev. 26 (2007) 82

[1]Historic England, UK
[2]Univ. Innsbruck, Austria

THE COMMON THREAD

Authenticating a Nazca tunic using combined ^{14}C and dye analysis

L. Hendriks[1], G.D. Smith[2], V.J. Chen[2], A. Holden[2], N. Haghipour

In 2020, the Indianapolis Museum of Art at Newfields (IMA) sought to acquire a Nazca dyed camelid wool tunic (Fig. 1), purportedly dated to the period 100 BCE – 600 CE. Because of the supple feel and excellent condition of the artifact, concerns were raised over its purported age. Museum curators and conservators requested an investigation of the object's materials.

Fig. 1: *a) Tunic, Nazca culture, 84 × 50 in., Roger G. Walcott Fund, 2021.177. b) Detail of pattern*

This report demonstrates for the first time the sequential, combined analysis of dyes by liquid chromatography-diode array detection-mass spectrometry (LC-DAD-MS) and subsequent ^{14}C dating of the same extracted yarns. Each of these techniques are destructive, requiring a small sample of the object, and as normally practiced are typically conducted separately by different laboratories on individual samples. Dye analysis requires a small sample in the order of hundreds of micrograms, while ^{14}C analysis consumes milligrams of material.

Of the wool samples removed from the tunic, all revealed the use of natural plant-based dyes. The analysis confirmed that the wool fibers were dyed with common Peruvian dyestuff (indigo blue, purpurin red, quercetin yellow) (see chromatogram in Fig. 2).

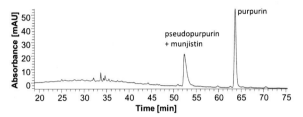

Fig. 2: *The chromatogram of the extracted red yarn shows major peaks whose DAD spectra and MS data are consistent with anthraquinones commonly found in Relbunium plant dyestuffs.*

The subsequent ^{14}C analysis of the extracted threads (≤0.5 mg) confirmed the purportedly Nazca attribution (100 BCE–600 CE, Fig. 3).

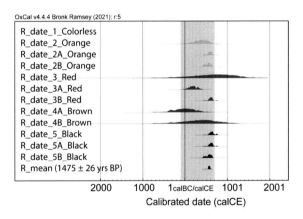

Fig. 3: *Radiocarbon age spread for the different measurements of the colored yarns.*

The mean value thereof provides yet a more definite timing of the object, as the data points to 550–650 CE, the Late Nazca period. Based on these confirmatory findings, the Nazca tunic was accessioned into the collection in 2021.

[1]   G. Smith et al, Herit. Sci. 10 (2022) 179

^{1}*HEIA, HESSO, Fribourg*
^{2}*Conservation Science Laboratory, Indianapolis Museum of Art at Newfields, USA*

8000 YEARS OF ALPINE MEADOWS USE

14C-dating on ibex, bears and sheep from the Gouffre de Giétroz

M. Blant[1], M. Luetscher[1], N. Reynaud Savioz[2], I. Hajdas

The Gouffre de Giétroz, discovered in 2017 on the south flank of the Susanfe Valley (Evionnaz, Valais, 2178 m alt), hosts many remarkably well-preserved bone-remains including Alpine ibex (*Capra ibex*), brown bears (*Ursus arctos*) and domestic sheep (*Ovis aries*). Field and laboratory work were carried out between 2017 and 2021 to sample, identify and analyze a rich corpus including more than 2000 bone fragments attributed to 16 mammal and bird species.

At least 15 ibex skulls (Fig. 1) as well as 15 to 18 skulls from sheep were found at the base of a 12 m deep shaft opening close to a steep cliff. The ibex population comprises exclusively large adult males. They belong to the original genetic stock of the Alpine region. Paleo-DNA studies carried out at the University of Zurich show that they form a homogeneous population. The sheep population includes primarily adult ewes and juvenile males. The 8 to 9 bears identified among the bone remains are all juveniles under 5 months old.

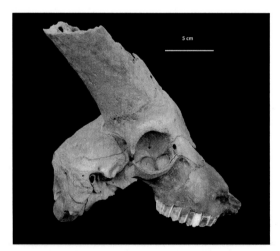

Fig. 1: *Ibex skull (EVG18/8, 6400-6246 BC).*

Radiocarbon dating was carried out on 18 skulls (11 ibex, 2 bears and 5 sheep) and 2 long bones (1 ibex and 1 sheep) to reconstruct the history of the alpine pasture, which is still being grazed by sheep nowadays. The 14C ages indicate that the cave acted as natural trap for nearly 8000 years (Fig. 2). The results reveal that ibex frequented the alpine grassland during the early Holocene. In particular, several individuals took advantage of the south-facing slop during the "8.2 ka event", a cold period known for its glacier advances. For several centuries, bears also had free access to the cave to hibernate.

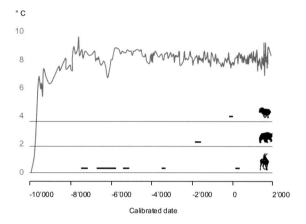

Fig. 2: *Chronological distribution of the main species identified in the shaft. Annual mean temperature curve after Affolter et al. 2019, Milandre cave [1].*

Pastoralism started during the "Roman warm period", c. 200 BC. Sheep were grazing in the area for at least a century and a half. The sheep were brought to the Giétroz alp during the Second Iron Age. This is the first evidence of ancient pastoralism at altitudes above 2000 m in the Valais Alps. To avoid too many losses, the Celtic shepherds built a wall to close a secondary entrance accessible from the cliff.

[1]   S. Affolter et al., Sci. Adv. 5 (2019) eaav3809

[1]*Swiss institute for Speleology and Karst studies, La Chaux-de-Fonds*
[2]*Laboratoire d'archéozoologie, Sion*

THE ROCK SHELTER OF FLÖZERBÄNDLI (MUOTATHAL, CH)

Epi-Palaeolithic Portable Art and Mesolithic Plant and Animal Remains

J.N. Haas[1], B. Dietre[1], F. Kleyhons[1], W. Kofler[1], W. Oberhuber[1], H. Stanger[1], I. Swidrak[1], W.H. Schoch[2], W. Imhof[3], W. Müller[4], I. Hajdas, U. Leuzinger[5]

Since 2006, an interdisciplinary research team aims at examining prehistorical sites in the Muotathal Alps (Ct. Schwyz, Switzerland) in archaeological and palaeoecological terms thanks to funding by the Staatsarchiv Schwyz. Many archaeological deposits yielded stone artefacts, charcoal, as well as plant and faunal remains. Radiocarbon analyses revealed more than 35000 year old cave bear remains, and a long-term human presence in the area since the Würmian Ice Age. Later, Epi-Palaeolithic (Azilien) portable art was found at Flözerbändli in form of a decorated piece of red deer antler (*Cervus elaphus*) with rows of bored pit marks, and which was radiocarbon dated to 10'273 ± 245 BC (cal. 2σ, ETH-109223), i.e. from the Younger Dryas chronozone [1].

Fig. 1: Left: Sediment dry sieving at Flözerbändli rock shelter during the archaeological excavation in 2021. Photo: J.N. Haas. Right: The decorated red deer antler fragment from the Flözerbändli rock shelter. Photo: W. Müller.

Also, Mesolithic camp sites of hunter-gatherers (9700 BC onwards), Neolithic to Roman hearths and bones from wild animals and livestock (7000 BC to AD 400), and numerous deserted Medieval and post-Medieval Alpine huts and enclosures are known from the area.

The Flözerbändli rock shelter is of great significance, as it also yielded Mesolithic sediments containing projectile points, animal bones, and well preserved charred fruits and seeds, as well as pollen due to extremely dry conditions in the rock shelter [1]. The Mesolithic hunter-gatherers burnt Scots pine, juniper, willow and rose in their fireplaces, and interestingly also yew wood (*Taxus baccata*) in greater amounts. Charred plant macrofossil finds comprised also alder and lime fruits, seeds of pinks and of St. John's wort (*Hypericum* cf. *perforatum;* potentially used for medical purposes), as well as hazelnut shells (*Corylus avellana*) collected and eaten as staple food.

Fig. 2: Charred hazelnut shell fragments from the Early Mesolithic layers. Photo: F. Kleyhons.

[1]   U. Leuzinger et al., Archäologisches Korrespondenzblatt 52(4) (2022) 461

[1]Botanik, Univ. Innsbruck, Austria
[2]Labor für quartäre Hölzer, Langnau am Albis
[3]Muotathal
[4]Lab. d'archéozoologie, Univ. Neuchâtel,
[5]Amt für Archäologie Thurgau, Frauenfeld

PRE-LGM LAKE SEDIMENTS IN THE SOUTHEASTERN ALPS

New radiocarbon dates from the Caltea Valley (NE-Italy)

L. Rettig[1], I. Hajdas, G. Monegato[2], P. Mozzi[1], M. Spagnolo[3]

Insights into the palaeoclimate and -vegetation of inner-Alpine valleys during interstadials preceding the Last Glacial Maximum (LGM) remain limited. This is partly because lake sediments from such warm periods have mostly been eroded by subsequent glacier advances. A notable exception can be found in the Caltea Valley, a narrow pre-Alpine valley in the Monte Cavallo Group (Venetian Prealps, NE-Italy). Here, a thick (ca. 8 m) sequence of laminated lacustrine deposits has been preserved, covered by fluvial gravels and ultimately by glacial till of the LGM ice advance (Fig. 1). The lake sediments contain a large number of plant macrofossils such as needles, cones, twigs, and even large pieces of tree trunk (Fig. 2), indicating the establishment of boreal forest vegetation at this site.

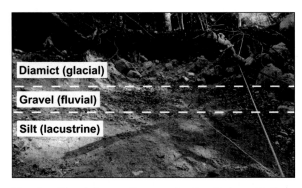

Fig. 1: *The lake sediments in the Caltea Valley, covered by fluvial gravels and till from the LGM glacier advance.*

The chronology of these sediments is poorly constrained with only a single radiocarbon date (29350 ± 460 ^{14}C a BP) being reported in the literature [1]. We have therefore collected a total of nine new samples from plant macrofossils in this section, which have been processed at LIP. Surprisingly, none of the samples yielded finite ^{14}C ages, even though different methods of preparation were applied. This indicates that the lake sediments date back beyond the limit of the radiocarbon method around 45 ka BP. They can potentially be attributed to the earliest parts of Marine Isotope Stage (MIS) 3, however, even an earlier interstadial (such as during MIS 5) cannot be ruled out with certainty.

Fig. 2: *Large tree trunk with characteristic post-depositional flattening, sampled for radiocarbon dating.*

Further examinations of the section, for example through more detailed pollen or macrofossil analyses may provide better chronological control. They may also reveal interesting insights into the palaeoclimate and -vegetation in the southeastern Alps during this pre-LGM interstadial. Our results also demonstrate that early radiocarbon chronologies may suffer from notable uncertainties, particularly if they are close to the age limit of the methodology.

[1]   F. Fuchs, Eiszeitalt. Ggw. 20 (1969) 68

[1]Geosciences, Univ. Padova, Italy
[2]CNR-IGG Padova, Italy
[3]Geosciences, Univ. Aberdeen, UK

RADIOCARBON DATING OF SKELETONS FROM BASEL

Double checking bone preparation for accurate chronology of burials

E. Flatscher,[1,2] M. A. Fortunato[1], I. Hajdas

A number of medieval and early modern burial grounds lie in the vicinity of the Historical Museum Basel (Basel, Switzerland) located in the former Franciscan church. A total of 265 burials north and south of the church (Fig. 1) have been excavated by the cantonal archaeological service (ABBS) in 2016-17 and 2020. These can be divided into one medieval phase north of the church (after ca. 1300), and three early modern phases south of the church (between 1580 and 1670).

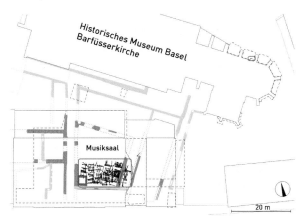

Fig. 1: *Excavated burial sites adjacent to the Historical Museum in Basel (marked in red). Plan: P. von Holzen, ABBS.*

A number of these burials can be dated unusually precisely through archaeological procedures: stratigraphy and artefacts contained in the graves, comparison with written sources and A-DNA pathogen analysis. Two skeletons could thereby be dated reliably to the last wave of the plague in Basel (1667-68).

For these reasons, samples from eight skeletons were selected for a collaboration project between the ABBS, the chair for Medieval Archaeology and Art History (UZH) and the Laboratory for Ion Beam Physics (ETH) to compare the results of archaeological and radiocarbon dating via two preparation methods.

From each of the selected individuals, two samples were taken from the ribs, one of which was prepared with the standard *Ultra Filtration method* (PREP1, base exposure time 20-30 min) and one with the base exposure time extended to 120 min (PREP2).

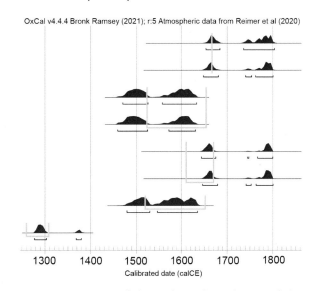

Fig. 2: *Average of the radiocarbon dating of the two samples from each individual skeleton, time span of archaeological dating in green. Graphic: P. von Holzen (ABBS).*

The differences between PREP1 and PREP2 have been statistically insignificant. As can be seen in fig. 2, the archaeological dating span and the 2 σ of the radiocarbon dating overlap in all cases. The project culminated in a bachelor thesis [1], a revised version is awaiting publication [2].

[1] M. A. Fortunato, 2021 unpublished
[2] E. Flatscher et al., Materialheft ABBS 33 in prep.

---

[1]*Medieval Archaeology, Univ. Zurich*
[2]*Archäologische Bodenforschung Basel-Stadt (ABBS)*

MODIS2 MORTAR DATING INTERCOMPARISON

Results of radiocarbon dating at ETH laboratory

I. Hajdas, A. Lindroos[1], Y. Gu[2]

Mortar has been used as a binder in building materials for thousands of years. From early human settlements, over 10 thousand years ago, up to the industrial production of modern cement (starting 19-20th century), whenever a lime mortar was used, the atmospheric CO_2 surrounding the buildings was fixed into its binder, setting the radiocarbon (^{14}C) clock for reading elapsed time from that moment. Separation of carbon for ^{14}C dating can be challenging because mortars are complex and heterogenous materials. Samples can be contaminated with unburned limestone or recrystallized calcite [1]. Multiple laboratories developed preparation methods to date archeological or historic sites. The first laboratory inter-comparison exercise was completed in 2016 [2].

The new inter-comparison was planned in 2019 and samples were distributed in the summer of 2020. Eleven laboratories took part in the study and results will be published in the upcoming proceeding of the 24th Radiocarbon and 10th Archaeology and Radiocarbon Conference, Zurich 2022.

Three samples (Tab. 1) were chosen for the study, prepared and distributed as a homogenized fine fraction (0-150 μm). In addition, laboratories received fragments of unprocessed mortar (Fig. 1).

Sample	Building	Location	Expected age
MODIS 2.1	church of Saltvik	Åland Islands, Finland	14th century
MODIS 2.2	church of Hamra	Swedish island of Gotland	14th century
MODIS 2.3	Basilica of Santa Eulalia in Mérida	western Spain	304-570 CE

Tab. 1: *Samples of mortar selected for the MODIS 2 inter-comparison.*

At the ETH laboratory, two associated wood and one charcoal fragments were analysed after being treated with acid-base-acid method. The fine powder was prepared by sequential dissolution. i.e., four fractions of CO_2 were collected in sequence of 3 seconds [3]. The results are in an agreement with the expected ages and it will be discussed in the upcoming publication.

[1] A. Ringbom et al., Radiocarbon 56 (2014) 619

[2] I. Hajdas et al., Radiocarbon 59 (2017) 1845

[3] I. Hajdas et al., Radiocarbon 62 (2020) 691

Fig. 1: *MODIS 2.1 mortar sample with a visible fragment of wood.*

[1]*Geology and Mineralogy, Åbo Akademi University, Finland*
[2]*Lab of AMS Dating, Nanjing University, China*

BOMB PEAK AND THE ANTHROPOCENE

Radiocarbon analysis for the definition of the GSPP

I. Hajdas, K. Wyss

Anthropogenic activities can be traced back to the early days of Homo sapiens. However, the impact on the Earth system, especially the emissions of greenhouse gases became evident after industrialization [1]. The International Commission on Stratigraphy (ICS) undertook first steps to investigate formalization of the Anthropocene as a geologic unit and in 2008 the Anthropocene Working Group AWG was formed (http://quaternary.stratigraphy.org/working-groups/anthropocene/). In 2019 the mid-20th century has been chosen by the AWG and was proposed to mark the onset of the Anthropocene epoch [2]. The following 2 years were dedicated to studies of sites for the definition of a global boundary strato-type section and point (GSSP). The research was founded by the House of the World Cultures (Haus der Kulturen der Welt, HKW) museum in Berlin and completed in 2022 [2].

The bomb peak ^{14}C is an excellent marker of the mid-20th century. Samples from six candidate sites (out of 12) were submitted to our laboratory (Tab. 1).

Site	Material	Ref.
East Gotland Baltic Sea	Sediment	[3]
West Flower Garden Bank	corals	[4]
Flinders Reef	corals	[5]
Crawford Lake, CA	sediment	[6]
Searsville Reservoir, USA	sediment	[7]
Śnieżka, Sudetes, PL	peat	[8]

Tab. 1: *Sites and samples type analysed at ETH. The results are either published or accepted for publication in The Anthropocene Review.*

Preparation of the samples was well documented and recorded for the purpose of the HKW project (Fig. 1).

Fig. 1: *Documentation of samples submitted as a part of the AWG and HKW project.*

The bomb ^{14}C was found in all but one record, which highly supported the chronology of the sites [3, 4, 5, 6, 8]. In the case of the Searsville Reservoir a high sedimentation of the od carbon from the catchment area obscured the bomb ^{14}C signal [7].

[1]　J. Zalasiewicz et al., Quat. Int. 383 (2015) 196

[2]　C. Waters and S. Turner, Science 378 (2022) 706

[3]　J. Kaiser et al., The Anthr. Rev. (2022) 20530196221132709

[4]　K. L. DeLong et al., The Anthr. Rev. (2022) submitted

[5]　J. Zinke et al., The Anthr. Rev. (2022) submitted

[6]　F. McCarthy et al., The Anthr. Rev. (2023) accepted

[7]　M.A. Stegner et al., The Anthr. Rev. (2023) 20530196221144098

[8]　B. Fiałkiewicz-Kozieł et al., The Anthr. Rev. (2022) 20530196221136425

[14]C ANALYSIS OF ATMOSPHERIC METHANE

Development of a new portable sampler

G. Zazzeri, L. Wacker, N. Haghipour[1], P. Gautschi, H. Graven[2]

The recent rapid growth in atmospheric methane (CH_4) is an immense challenge to society, as CH_4 is the second most important anthropogenic greenhouse gas. Although there are several competing hypotheses, we still don't know why methane is rising [1] (Fig. 1).

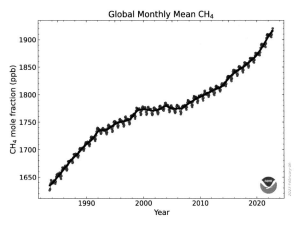

Fig. 1: *Global trend of atmospheric CH_4 from NOAA Global monitoring laboratory.*

[14]C analysis offers a powerful tool to understand methane emissions because it enables distinction between fossil (natural gas) vs biogenic methane sources (agriculture, landfill sites). Fossil-derived CH_4 is entirely devoid of ^{14}C; when emitted into the atmosphere, it strongly decreases the atmospheric ratio of ^{14}C to total carbon in CH_4 ($\Delta^{14}CH_4$). By observing changes in $\Delta^{14}CH_4$, the fossil fraction of CH_4 emissions can be quantified. However, ^{14}C measurements of atmospheric CH_4 are particularly challenging, because CH_4 concentrations are low (~1.9 ppm) and a large amount of air must be sampled in order to collect enough carbon for ^{14}C analysis.

At LIP we developed a portable sampler based on a prototype [2]. The new system enables extraction of carbon from CH_4 while sampling in the field, reducing the sample processing in the laboratory and allowing collection of a hundred liters of air onto a 0.5 g zeolite trap.

The system is based on three main steps: 1) H_2O, CO and CO_2 removal, 2) combustion of CH_4, 3) adsorption of the combustion-derived CO_2 onto a zeolite sample trap (Fig. 2).

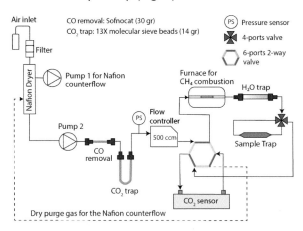

Fig. 2: *Sampling system schematic. Dark green lines in the 6-ports valve indicate the flow direction. In this configuration, the CO_2 sensor measures CO_2 concentrations after the sample trap, to check for the trap breakthrough. By switching the 6-ports valve, CO_2 is measured before the furnace, to check that all the CO_2 has been trapped.*

After each sampling, the sample trap is disconnected from the system. The combustion-derived CO_2 is then desorbed by heating the sample trap at 400 °C, cryogenically trapped and sealed into a glass ampule to be used in the gas interface of the AMS system Micadas.

[1] E. G. Nisbet et al., Rev. Geophys. 58.1 (2020) e2019RG000675

[2] G. Zazzeri et al., Environ. Sci. Technol. 55.13 (2021) 8535

[1]*Earth Sciences, ETH Zurich*
[2]*Physics, Imperial College London, UK*

RICH: DISSOLVED CARBON SPECIES IN SWISS LAKES

Monthly measurements of ^{14}C in Lake Constance and Lake Geneva

M. White[1], B.V.A. Mittelbach[1], T. Rhyner[1], T.M. Blattmann[1], N. Haghipour, M. Wessels[2], N. Dubois[3], T.I. Eglinton[1]

Radiocarbon Inventories of Switzerland (RICH) aims to construct the first national-scale census of carbon across aquatic, terrestrial, and atmospheric reservoirs. Within the carbon cycle, inland waters play a crucial role with lakes integrating carbon from various sources within their catchment in addition to that fixed by local primary productivity. We made measurements of DO^{14}C and DI^{14}C from monthly water column samplings of Switzerland's two largest lakes- Lake Constance and Lake Geneva.

Preliminary results show that the average radiocarbon signature of DIC in both lakes is depleted relative to atmospheric CO_2, indicating around 15% contribution from limestone-derived (^{14}C-dead) carbon. The timeseries at Lake Constance builds on earlier measurements which have showed a decrease in DI^{14}C since the late 1960s [1,2] due to decreasing concentrations of bomb radiocarbon in the atmosphere. Measurements of DO^{14}C in Lake Constance are more enriched compared to DI^{14}C, indicating the presence of terrestrial DOC.

Results from Lake Geneva indicate significant river influence at the sampling site (LeXPLORE platform), which appears to deliver ^{14}C-depleted DIC to the lake. The extent of this influence, and the composition of the river plume, vary seasonally. Measurements of DO^{14}C in Lake Geneva are similar to DI^{14}C, consistent with lake primary productivity as the main source of DOC.

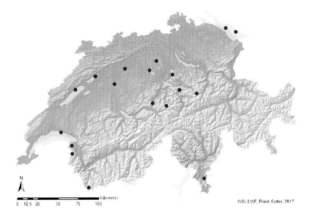

Fig. 2: *Lakes from across Switzerland have been sampled for radiocarbon measurement of DIC and DOC.*

Future work will complete the yearlong timeseries of DI^{14}C and DO^{14}C at Lake Constance and Lake Geneva. In addition, approximately 15 other lakes around Switzerland, covering a range of sizes, catchment characteristics, and eutrophic states have already been sampled and will be measured for radiocarbon at LIP in the coming year (Fig. 2).

[1] T. M. Blattmann et al., Radiocarbon 791 (2018) 60

[2] W. Kölle, Vom Wasser 34 (1969) 36

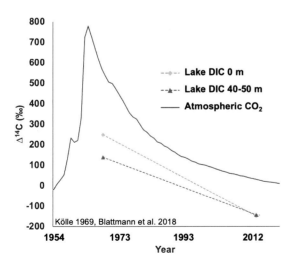

Fig. 1: *The radiocarbon signature of dissolved inorganic carbon (DIC) in Lake Constance has decreased since the 1960s [1, 2].*

[1]*Environmental Systems Sciences, ETH Zurich*
[2]*Seenforschung, LUBW, Langenargen, Germany*
[3]*Eawag, Dübendorf*

ORGANIC CARBON SOURCES OF LAKE CONSTANCE

Using sedimentary ^{14}C signatures to trace the origin of organic matter

B.V.A. Mittelbach[1], M.E. White[1], T.M. Blattmann[1], N. Haghipour, M. Wessels[2], N. Dubois[3], T.I. Eglinton[1]

Organic carbon burial in lakes removes carbon from Earth's surface pools, but the effect on the atmosphere depends on the type of C. The sequestration of biospheric OC represents a CO_2 drawdown, but rock-derived OC exerts no net effect on atmospheric CO_2 levels. Here, we combine Δ^{14}C and stable δ^{13}C signatures of bulk sedimentary OC to constrain the nature and dynamics of OC accumulation in perialpine Lake Constance.

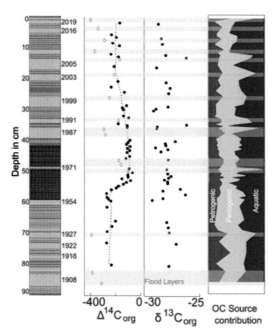

Fig. 1: *The sediment core was characterized by flood layers (grey) and varved sedimentation.*

Bedrock and soil samples from the lake catchment and aquatic biomass from a sediment trap were collected [1]. Δ^{14}C was modeled for soil carbon using a turnover-time of 67 ± 23 years [2] and lake DI^{14}C was modeled using a turnover time of 5 yr with DIC$_{Atm}$ content of $80 \pm 5\%$ [3]. Bedrock was assumed to be radiocarbon free. OC was assumed to stem from one of these three pools. We used a linear mixing model to calculate the fractional contribution for each core subsample.

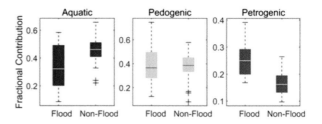

Fig. 2: *Flood deposits have variable aquatic and soil-derived C, depending on the source of the flood.*

Analysis at quasi-annual resolution shows a muted but distinct bomb spike with Δ^{14}C values of bulk organic carbon (OC) increasing from -250‰ to -100‰ in the early 1960s. Our linear mixing model reveals that aquatic biomass and soil-derived, pre-aged OC comprise the largest contributors to the sedimentary OC pool. We attribute the presence of the bomb spike signal to the rapid incorporation of bomb-derived ^{14}C into the lake DIC pool. While TOC content increased during the mid-century eutrophication interval, no carbon pool seems to be preferentially preserved. Flood layers are characterized by a higher relative contribution of soil and rock-derived carbon, leading to their older ^{14}C signature. We interpret the large spread of aquatic and soil-derived C in the floods to reflect different flooding mechanisms, such as overland flow or sediment remobilization.

[1] T. M. Blattmann et al., Chem. Geol. 503 (2019) 52

[2] T. S. van der Voort et al., Biogeosci. 13 (11) (2016) 3427

[3] T. M. Blattmann et al., Radiocarbon 60 (3) (2018) 791

[1]Earth Sciences, ETH Zurich
[2]ISF, Langenargen, Germany
[3]Eawag, Dübendorf

WHAT FACTORS ARE DRIVING THE SOIL CARBON CYCLE?

A radiocarbon inventory of soils in Switzerland

M. Moreno Duborgel[1,2], L.I. Minich[1,2], N. Haghipour, T.I. Eglinton[2], F. Hagedorn[1]

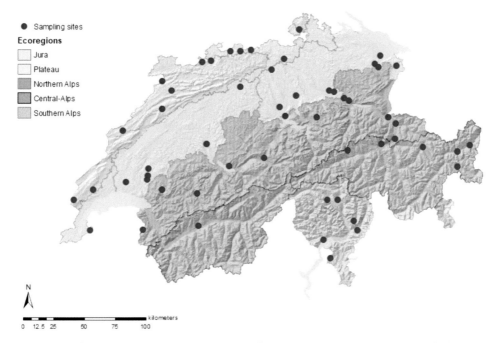

Fig. 1: Map of the 54 forest sites where mineral soil from 0-20 cm depth was sampled.

Soils are the largest reservoir of carbon in terrestrial ecosystems. However, global warming rises numerous uncertainties regarding the fate of soil organic carbon and the role that soils will play in carbon storage. This study is part of the radiocarbon inventory of Switzerland project. This inventory aims at reconstructing the carbon cycle across Switzerland and quantifying the carbon fluxes between the different reservoirs (i.e. atmosphere, aquatic and terrestrial ecosystems). In this study, we are focusing on soils. How long is the carbon staying in Swiss soils? What are the factors driving carbon turnover times?

We are using radiocarbon to calculate carbon turnover times from soils across 54 forest sites in Switzerland (Fig. 1). These sites span a broad range of climate and geological conditions. Mineral soil from these 54 forest sites, from 0 to 20 cm depth, was sampled repeatedly in 1990, 2014 and 2022. To better understand carbon

dynamics in soils, the mineral soil was separated into particulate organic matter (POM) and mineral associated organic matter (MAOM) according to density. Our results show that the POM has a turnover time on a decadal time scale whereas the MAOM turns over on a centennial to millennial timescale. Early results suggest that the stabilization of the MAOM is mostly driven by soil properties. For example, in soils with low pH, organic matter is bound to pedogenic oxides. In soils with high pH, calcium cations are stabilizing the organic matter [1]. The particulate organic matter is more prone to climatic variables such as precipitation and temperature.

[1] M. C. Rowley et al., Biogeochem. 137 (2018) 27

[1]*Swiss Federal Institute for Forest, Snow and Landscape Research (WSL), Birmensdorf*
[2]*Earth Sciences, ETH Zurich*

MAGNITUDE AND SOURCES OF CARBON FLUXES

A radiocarbon inventory of carbon fluxes from soils in Switzerland

L.I. Minich[1,2], M. Moreno Duborgel[1,2], N. Haghipour, T.I. Eglinton[2], F. Hagedorn[1]

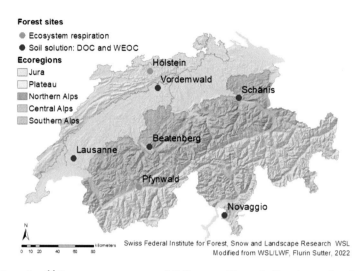

Fig. 1: *Swiss forest sites for ^{14}C measurements of C fluxes. Blue: Collection of soil solution for DO^{14}C and WEO^{14}C measurements. Green: Collection of soil respired CO_2.*

Soils are the largest terrestrial carbon (C) reservoir with an estimated Swiss C stock of 450 Mt C. Soils exchange C vertically with the atmosphere as carbon dioxide (CO_2) and laterally with the aquatic C pools as dissolved organic (DOC) and inorganic (DIC) carbon. Climate change induces perturbations in the carbon cycle, likely increasing C fluxes from soils of hitherto stabilized C to the atmosphere and aquatic domain. However, the exchange of C between the different C reservoirs is one of the greatest uncertainties in the C cycle and its feedback on climate change. This study is part of the RICH (Radiocarbon Inventories of Switzerland) project, aiming at better understanding the magnitude and sources of C fluxes from soils to the atmosphere and aquatic C pools (i.e. rivers, lakes). We use ^{14}C measurements of DOC in soil solution and water extractable organic carbon (WEOC) from five Swiss forest sites (Fig. 1) to explore C transfer from topsoils into subsoils and further C export to aquatic systems [1]. We further use ^{14}C data of soil respired CO_2 from different land use types to elucidate differences in sources of respired CO_2 across these land use types. In addition, we compare respired $^{14}CO_2$ from two forest soils subjected to different soil moisture conditions to reveal variability of C flux rates and source contribution. Preliminary data revealed small DOC fluxes from soils to aquatic ecosystems, a strong decrease in DOC concentration with a concurrent increase in the ^{14}C age of DOC with soil depth. This is likely related to the interaction of DOC with soil minerals: "young", leached DOC is sorbed to minerals and remobilized during microbial processing, then contributing to an "older" DOC source. Site-comparison data further suggest that spatial variation of DO^{14}C may rather be driven by physico-chemical differences in soils than by climatic differences across sites. Our findings indicate the importance of DOC for carbon stabilization in soil.

[1] T. S. van der Voort, et al., Biogeosciences. 16 (2019) 3233

[1]*Swiss Federal Institute for Forest, Snow and Landscape Research (WSL), Birmensdorf*
[2]*Department of Earth Sciences, ETH Zurich*

SPATIAL MACHINE LEARNING USING MOSAIC DATABASE

Predicting the distribution of radiocarbon in the East Asian margin

S. Paradis[1], M. Diesing[2], H. Gies[1], N. Haghipour, T.I. Eglinton[1]

With the expansion of available data from marine geosciences studies, there is a growing need to harmonize and compile data of organic carbon (OC) and associated parameters to understand the role of marine sediments in the global carbon cycle. Hence, the Modern Ocean Sediment Archive and Inventory of Carbon (MOSAIC) database was devised [1], centered on compiling data of radiocarbon in marine sediments as a powerful tracer and chronometer of carbon cycle processes. This database currently holds > 2000 datapoints of radiocarbon measurements in surficial marine sediment which can be used to train spatial machine learning models to understand its distribution in the marine realm.

Given the wealth of data in the East Asian margin (n > 1000; Fig. 1) and complex processes that provide a wide range of ^{14}C values, this margin proved an ideal case study to predict the distribution of radiocarbon and other geochemical properties (OC, total nitrogen, δ^{13}C, and surface area, data not shown).

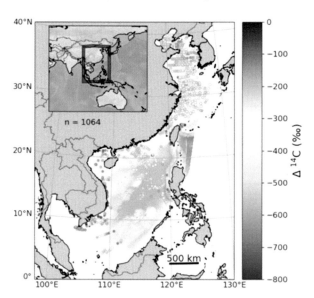

Fig. 1: *Distribution of surficial Δ^{14}C in the East Asian margin.*

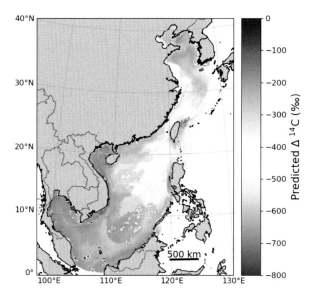

Fig. 2: *Spatial prediction of surficial Δ^{14}C.*

The prediction surface of radiocarbon shows that marine sediments generally consist of young OC (> -100‰). Highly depleted Δ^{14}C values (-800‰) were observed surrounding Taiwan, given the erosive nature of mountainous rivers that discharge significant petrogenic OC to the margin. Aged OC was also predicted offshore the Chiangjiang river mouth, due to the discharge of degraded and pre-aged terrestrial OC by this river in comparison to other rivers. Similarly, low Δ^{14}C values were observed in the deep South China Sea due to the long transit time prior to its deposition, which promotes the degradation and ageing of OC.

This case study is an example of how large datasets can be used to understand the carbon cycle processes in marine sediments.

[1] T. S. Van der Voort et al., Earth Syst. Sci. Data 13 (2021) 2135

[1]Earth Sciences, ETH Zurich
[2]Geological Survey of Norway, Trondheim, Norway

PERMAFROST CARBON ON THE BEAUFORT SHELF

A multi-disciplinary effort to investigate a rapidly changing Arctic Sea

L. Bröder[1,2], J. Lattaud[1], B. Juhls[3], A. Eulenburg[3], T. Priest[4], M. Fritz[3], A. Matsuoka[5], A. Pellerin[6], T. Bossé-Demers[7], D. Rudbäck[8], M. O'Regan[8], D. Whalen[9], N. Haghipour, T.I. Eglinton[1], P. Overduin[3], J. Vonk[2]

Arctic continental shelves are profoundly impacted by rising air temperatures and declining summer sea-ice extent. Coasts are rapidly eroding and subsea permafrost thawing, river runoff is warming and affecting associated particulate and dissolved matter fluxes, with direct consequences for the marine environment and profound ramifications for the broader Earth climate system by impacting carbon turnover, ocean acidification, and greenhouse gas fluxes between sediment, sea and atmosphere (Fig .1).

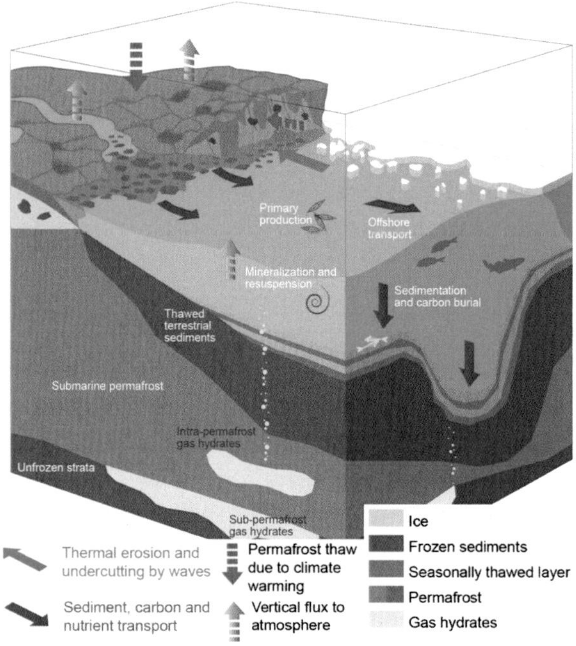

sediment sampling were conducted along five major across-shelf transects (Fig. 2) in order to quantify the fluxes, burial rates, composition and fate of organic matter to ultimately improve assessments of the Beaufort shelf as a carbon source or sink, and place these outcomes in the context of the Holocene paleoenvironment and transgressed permafrost. Radiocarbon analyses help to constrain organic matter source contributions (e.g. radiocarbon-dead petrogenic vs modern marine) and to quantify carbon burial rates (via age-depth relations for longer sediment cores).

Fig. 1: *Scheme of the various processes affecting Arctic continental shelves.*

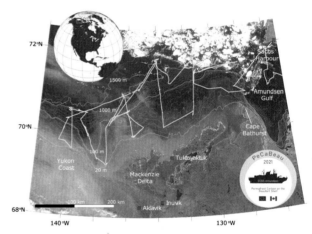

Fig. 2: *Cruise track of the PeCaBeau expedition on CCGS Amundsen in 2021.*

[1] L. Bröder et al., Reports on Polar and Marine Research 759 (2022)

The multi-disciplinary PeCaBeau project ("Permafrost Carbon on the Beaufort Shelf") aims to track the movement and transformation of material from permafrost thaw along the land-to-ocean continuum. Sampling operations took place in the southern Beaufort Sea in September 2021 onboard the Canadian Coast Guard Ship *Amundsen* [1]. Water-column profiling and

[1]Earth Sciences, ETH Zurich
[2]Vrije Univ. Amsterdam, Netherlands
[3]AWI Potsdam, Germany
[4]MPI for Marine Microbiology, Bremen, Germany
[5]Univ. New Hampshire, Durham, USA
[6]Univ.du Québec à Rimouski, Canada
[7]Univ. Laval, Canada
[8]Stockholm Univ., Sweden
[9]Natural Resources Canada

ORGANIC CARBON IN THE DEEP SARGASSO SEA

Tracing bomb radiocarbon in sinking particulate

C. Schnepper[1], R. Pedrosa-Pàmies[2,3], M. Conte[3], N. Gruber[4], N. Haghipour, T.I. Eglinton[1]

The imprint of bomb radiocarbon in sinking particulate organic carbon (PO^{14}C) intercepted by sediment traps, together with flux and compositional data, provides information about the origin and dynamics of oceanic particles [1]. To quantify the intra- and inter-annual variability in PO^{14}C, an in-depth study was initiated at the Ocean Flux Program site in the Sargasso Sea.

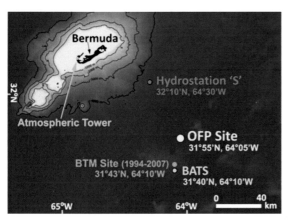

Fig. 1: *Map of Bermuda and the surrounding sampling stations (Weber, JC, 2022).*

First POC measurements from 2012, 2013 and 2014 show a decrease in Δ^{14}C with depth and a range in Δ^{14}C of 80‰ for 500 m, 52‰ for 1500 m and 73‰ for 3200 m. The large amplitude in the shallow trap already indicates that sinking PO^{14}C from the Sargasso Sea is more depleted than expected just through vertical settling from the surface and that particles from other, older carbon sources dilute the ^{14}C content. The enriched Δ^{14}C in the 3200 m trap resembling the bomb spike signature on the other hand suggests a rapid settling of younger organic matter being transported to the deep sea.

Analysis at quasi-annual resolution shows a muted, but distinct, bomb spike with Δ^{14}C values of bulk organic carbon (OC) increasing from -250‰ to -100‰ in the early 1960s. Our linear mixing model reveals that aquatic biomass and soil-derived, pre-aged OC comprise the largest contributors to the sedimentary OC pool.

A rapid organic matter remineralization in the water column reduces the fraction in the mesopelagic zone. In the bathypelagic zone, remineralization is less effective.

Fig. 2: *The 3200 m OFP Sediment trap, October cruise 2022.*

The influence of lithogenic material on the flux composition is reversed compared to that of organic matter, with a higher proportion and lower Δ^{14}C in deeper sediment traps. This would support the hypothesis that a substantial part of the deep-sea sinking particle flux is deriving from lateral transport or particle resuspension [2].

[1]   J. Hwang et al, Global Biogeochem. Cycles 24 (2010) GB4016

[2]   M. H. Conte et al., Chem. Geol. 511 (2019) 279

[1]*Earth Sciences, ETH Zurich*
[2]*Bermuda Institute of Ocean Sciences, St. Georges, Bermuda*
[3]*Marine Biological Laboratory, Woods Hole, USA*
[4]*Environmental Systems Sciences, ETH Zurich*

COUPLED ^{14}C AND ^{13}C MEASUREMENTS OF SPA-127

Comparison of novel and established measurement techniques

M. Wertnik[1], S. Bernasconi[1], M. Jaggi[1], L. Wacker, N. Haghipour[1], C. Welte[1]

Stable carbon has been routinely measured in stalagmites as it is easily collected alongside oxygen isotope measurements which are an important staple in speleothem research. However, interpreting the stable carbon signature can be quite difficult on its own, as the final δ^{13}C value of the stalagmite is composed of various sources of carbon: from vegetation to organic matter pools in the soil and even the Karst bedrock itself. Having an accompanying radiocarbon signature to complement the stable carbon signature can alleviate this problem by allowing to exclude certain sources of carbon.

We developed a method for simultaneous carbon isotope measurements of carbonates using the gas source interface (GIS) [1]. This GIS-AMS IRMS method was applied to a stalagmite with high variability in both isotopes. For this stalagmite, SPA-127 from Spannagel cave in Austria (Fig. 1), both a ^{14}C record from laser ablation (LA) AMS as well as a high resolution δ^{13}C record exist [2].

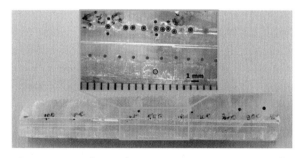

Fig. 1: *sampling locations on SPA-127 for Coupled (magenta circle), original (teal dots) and new (purple dots) conventional IRMS measurements. The white lines visible along the stalagmite are the laser tracks from LA-AMS.*

In Fig. 2, the results of the measurements with the novel GIS-AMS-IRMS method of SPA-127 are shown and compared to both existing and new data from conventional measurements.

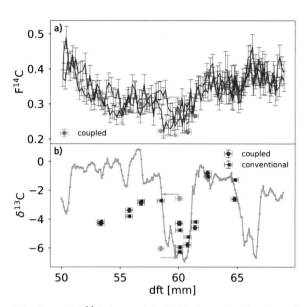

Fig. 2: *a) F^{14}C from LA-AMS (purple line) and coupled GIS-AMS-IRMS (orange circles), b) δ^{13}C from original (teal line) and new (purple triangles) conventional IRMS, and coupled GIS-AMS-IRMS (magenta circles), vs. distance from top (dft) in mm. Labels of grey samples were most likely mixed up.*

While the F^{14}C agrees within uncertainties with the LA-AMS data, for the δ^{13}C the coupled measurements had a deviation of up to 4‰ from the original conventionally measured data. However, further conventional measurements at the same locations agree with the coupled data, suggesting that we see a real difference in δ^{13}C signature between the sampling locations closer and further away from the growth axis of the stalagmite (Fig. 1).

[1]   M. Wertnik et al., Radiocarbon in prep.
[2]   C. Welte et al., Clim. Past, 17 (2021) 2165

[1]Earth Sciences, ETH Zurich

RADIOCARBON ANALYSIS OF FOSSIL POLLEN

Flow cytometry particle sorting for AMS dating

K. Nakajima, C. Heusser, C. Welte, L. Wacker, T. I. Eglinton[1]

Current AMS technology allows ^{14}C analysis of samples containing just few µg of carbon. This has facilitated novel applications of ^{14}C, such as compound- or particle-specific ^{14}C analysis, which previously were limited by sample size. One such application is the use of flow cytometry, a particle separation technique well established in clinical research, for separating optically characteristic particles from environmental matrices for subsequent ^{14}C analysis, e.g. [1]. Within the framework of a PhD project, this approach is utilized using a dedicated flow cytometer located at the ETH Biogeoscience Group for ^{14}C dating of fossil pollen recovered from sediment cores.

Fossil pollen are ubiquitously available and well preserved in sediments, and they are established markers for past ecosystem variabilities. Thus, pollen present a suitable target for generating proxy- and source-specific ^{14}C chronologies that are not limited by availability compared to traditional targets such as macrofossils. Pollen are extracted from sediments through a series of physicochemical steps with flow cytometry as the final separation step before transfer to tin capsules for combustion. Sufficient material can be obtained from 1-2 g dry sediment for EA-AMS.

Blank assessment on modern and ^{14}C-dead pollen material based on the method of constant contamination [2] suggest that the introduction of extraneous carbon from processing to the final sample is marginal (Fig. 1). Thus, corrections for samples ranging from 50-100 µg C, a typical sample size with 100k pollen, are rather small.

Pollen-^{14}C dates were investigated for two lakes for layers with independent age estimates, either from macrofossil-[14]C [3] or tephrochronology [4]. Pollen-^{14}C dates were coherent with all macrofossil and tephra dates, whereas other organic particle fractions partly deviated significantly. These results suggest that this approach provides

a chronological tool that is a viable alternative to existing methods. The project is now moving onto routine application of this method.

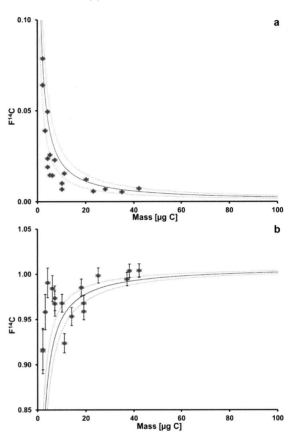

Fig. 1: *Blank assessment for tracing a) modern contamination with ^{14}C-dead pollen material and b) fossil contamination with modern pollen, which constitute 0.85 ± 0.21 µg C extraneous carbon of F[14]C = 0.27 ± 0.07 from processing.*

[1] R. K. Tanner et al., J. Quat. Sci. 28 (2013) 229

[2] U. M. Hanke et al., Radiocarbon 59 (2017) 1103

[3] S. Engels, Boreas 50 (2021) 519

[4] G. Jones et al., Boreas 47 (2021) 519

[1]*Earth Sciences, ETH Zurich*

COSMOGENIC NUCLIDES

Drone view (N. Akçar) of Fanaus outcrop at Tamins rock avalanche

(Sub)glacial landforms of the LGM Rhine glacier

The Last Glacial Maximum of Ticino-Toce Glacier

Last Glacial Maximum (LGM) ice conveyor belts

The Last Glacial Maximum in the Orobic Alps

Deglaciation of the Alps and the Lateglacial

Postglacial denudation in the Dora Baltea Valley

Quantifying post-glacier bedrock erosion rates

Using P-PINI to decode the Deckenschotter

Early Pleistocene stratigraphy in the Alps

Reconstructing the Tamins rock avalanche

LGM glacial advances in Barhal Valey (NE, Türkiye)

^{10}Be in Black Sea sediments during Termination II

Internal deformation of the Anatolian Scholle

Controls on river incision and aggradation

Synchronous records of past Baltic change

Drill chips from Little Dome C and solar activity

The ^{10}Be/^{36}Cl ratio as a dating tool

The solar signal in ice core excess water samples

(SUB)GLACIAL LANDFORMS OF THE LGM RHINE GLACIER

High-resolution geomorphological mapping reveals landform patterns

S. Kamleitner, S. Ivy-Ochs, B. Salcher[1], J.M. Reitner[2]

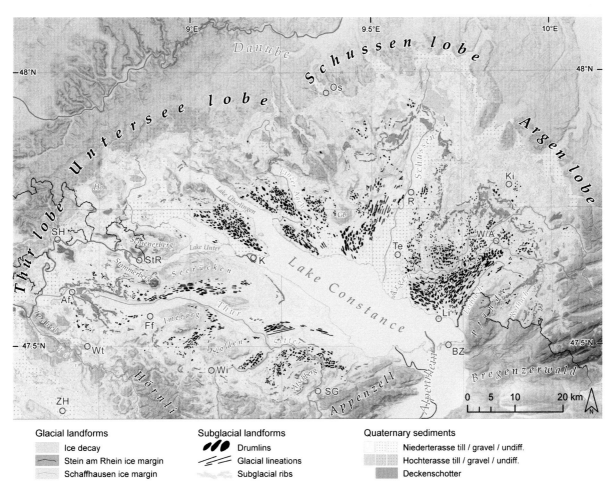

Fig. 1: *Glacial and subglacial landforms mapped within the footprint of the LGM Rhine glacier, Swiss-German Alpine foreland [1, 2].*

Using geomorphological mapping on the basis of LiDAR elevation data, we visualize the wealth of (sub)glacial landforms within the footprint of the Last Glacial Maximum (LGM) Rhine glacier (Fig. 1), one of the largest piedmont glaciers of the Alps. We thereby generate a digital landform inventory on piedmont lobe scale that provides key insights for reconstructing extent and flow of Rhine paleoglacier [1, 2]. Next to two sets of ice margins (Schaffhausen, Stein am Rhein), each comprised of multiple lateral and frontal moraine ridges, more than 2500 subglacial features were mapped. Drumlins make up the majority of streamlined bedforms (n=2460) and are typically organized in fields aligned in a scalloped pattern between the shore of Lake Constance and the Stein am Rhein ice margin. Glacial lineations and subglacial ribs were identified additionally.

[1] S. Kamleitner et al., Geomorphology 423 (2023) 108548

[2] S. Kamleitner, Diss. ETH No. 28443

[1]*Geography and Geology, Univ. of Salzburg, Austria*
[2]*Geological Survey of Austria, Vienna, Austria*

THE LAST GLACIAL MAXIMUM OF TICINO-TOCE GLACIER

A new glacier chronology based on geomorphology and exposure dating

S. Kamleitner, S. Ivy-Ochs, G. Monegato[1], F. Gianotti[2], N. Akçar[3], C. Vockenhuber, M. Christl, H.-A. Synal

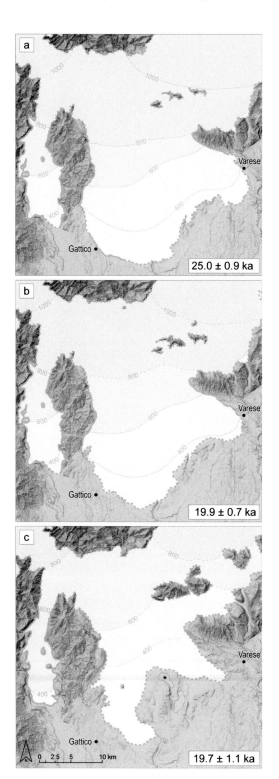

We present a detailed reconstruction of the chronological and spatial evolution of the Ticino-Toce glacier throughout the peak of the Last Glacial Maximum (LGM) based on geomorphological landform analysis combined with new 41 [10]Be and [36]Cl erratic boulder surface exposure ages.

Ticino-Toce glacier reached its LGM maximum extent at ca. 25.0 ± 0.9 ka [1]. At that time the Verbano lobe covered about 380 km^2 of the northern Italian foreland, forming a wide two-lobed piedmont glacier (Fig. 1a). In contrast, the LGM extent of the smaller Orta lobe to the west (85 km^2) remained topographically more strongly constrained [2]. Over the next ~5000 years, the Ticino-Toce glacier system underwent several minor oscillations during which the glacier front repeatedly restabilized close to its LGM maximum position (Fig. 1b). After 19.9 ± 0.7 ka, the ice surface level dropped significantly [1]. Although interrupted by a glacier readvance at ca. 19.7 ± 1.1 ka (Fig. 1c), collapse of the foreland glacier must have occurred shortly after [1]. The new chronology of the Ticino-Toce glacier system matches temporal constraints of other glaciers outflowing the Central Alps well, yet highlights differences in glacier behaviour to eastern Southern Alpine glaciers, possibly linked to shifting LGM precipitation patterns.

Fig. 1: *Paleoglaciological reconstruction of the LGM Ticino-Toce glacier system with Orta (west) and Verbano (east) piedmont lobes (after [1, 2]).*

[1]   S. Kamleitner et al., Quat. Sci. Rev. 279 (2022) 107400
[2]   J. Braakhekke et al., Boreas 49 (2020) 315

[1]*Geosciences and Earth Resources, CNR, Padua, Italy*
[2]*Earth Sciences, Univ. degli Studi di Torino, Italy*
[3]*Geology, Univ. of Bern*

LAST GLACIAL MAXIMUM (LGM) ICE CONVEYOR BELTS

Boulder trajectories of the LGM Ticino-Toce glacier network

G. Monegato[1], S. Kamleitner, F. Gianotti[2], S. Martin[3], C. Scapozza[4], S. Ivy-Ochs

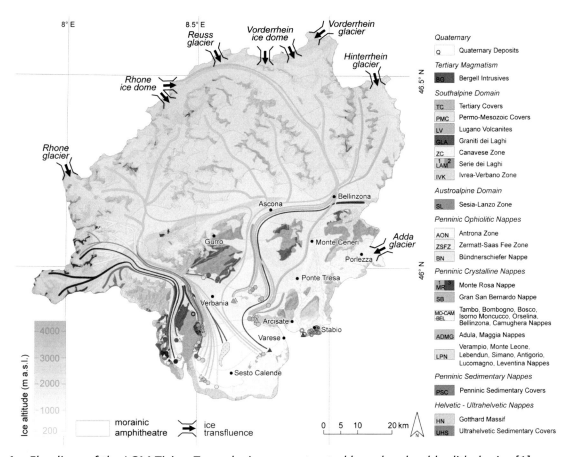

Fig. 1: *Flowlines of the LGM Ticino-Toce glacier reconstructed based on boulder lithologies [1].*

The provenance and distribution of erratic boulders within the Orta and Verbano end moraine systems provide exciting insights into organisation and dynamics of the Ticino-Toce glacial network at the LGM, highlighting the importance of topography in driving differential glacier evolution.

Erratic boulders located in the Orta basin as well as the western-central sector of the Verbano end moraine system were found to be sourced from the Toce catchment. On the contrary, lithologies indicative of the Ticino subcatchment are limited to the eastern sector of the former Verbano lobe. According to this pattern, we suggest that ice of the Toce glacier not only fed the Orta lobe but extended far into the southeast of the Verbano lobe (Fig. 1). Despite the larger catchment, LGM Ticino glacier was limited to the eastern flank of the Verbano system. An early arrival of LGM Toce glacier on the piedmont plain, promoted by the high and steep slopes of the Toce catchment, may have led to damming of Ticino glacier and fostered the diffluence of Ticino ice into the prealpine area to the east.

[1]   G. Monegato et al., AMQ 35 (2022) 119

[1]*Geosciences and Earth Resources, CNR, Padua, Italy*
[2]*Earth Sciences, Univ. degli Studi di Torino, Italy*
[3]*Geosciences, Univ. of Padua, Italy*
[4]*Earth Science, SUPSI, Canobbio*

THE LAST GLACIAL MAXIMUM IN THE OROBIC ALPS

New [36]Cl exposure ages from the Valle d'Inferno (Northern Italy)

L. Rettig[1], S. Kamleitner, S. Ivy-Ochs, P. Mozzi[1], G. Monegato[2], M. Spagnolo[3]

The chronology of the Last Glacial Maximum (LGM) in the southern Central Alps is largely based on studies of the Verbano and Garda amphitheaters [1, 2]. In between those large glacier lobes, in the Orobic Alps, isolated glaciers developed that were confined to smaller valleys or cirques (Fig. 1). The reconstruction of these glaciers and their Equilibrium Line Altitudes (ELAs) can provide valuable insights into the LGM paleoclimate in the Southern Alps. However, numerical dating of moraines in the Orobic Alps has never been attempted. It therefore remains unclear, if these moraines represent an LGM advance and if so, whether the smaller glaciers in the Orobic Alps advanced synchronously with the Verbano and Garda lobes.

Fig. 2: *The lateral-frontal moraine complex of the Valle d'Inferno. Crestlines are marked with red arrows.*

Our results demonstrate that the moraines in the Valle d'Inferno were indeed formed during the global LGM, synchronous to the maximum advance of larger outlet glaciers in the southern Central Alps. The new dates provide the necessary age control for paleoclimatic reconstructions that will be attempted in the next stages of this project.

[1] S. Kamleitner et al., Quat. Sci. Rev. 279 (2022) 107400
[2] G. Monegato et al., Sci. Rep. 7 (2017) 2078
[3] J. Ehlers and P. L. Gibbard, Quaternary Glaciations - Extent and Chronology (2004)

[1]Geosciences, Univ. Padova, Italy
[2]CNR-IGG Padova, Italy
[3]Geosciences, Univ. Aberdeen, UK

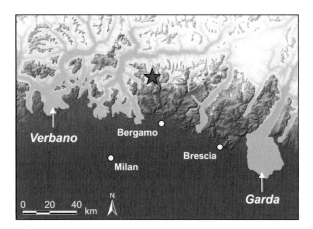

Fig. 1: *The southern fringe of the Central Alps during the LGM. Ice extent according to [3]. The location of the Valle d'Inferno is marked with a red star.*

One of these areas of isolated glaciation is the Valle d'Inferno in the western part of the Orobic Alps (Fig. 1). A pair of very prominent lateral to frontal moraine ridges indicates the extent of the former glacier that occupied this valley (Fig. 2). The crestlines of the moraines are partially covered with large, glacially transported boulders, six of which were sampled for [36]Cl exposure dating at LIP.

DEGLACIATION OF THE ALPS AND THE LATEGLACIAL

European Glacial Landscapes: the Gschnitz and Egesen stadials

S. Ivy-Ochs, G. Monegato[1], J.M. Reitner[2]

As the foreland piedmont glaciers collapsed after the Last Glacial Maximum (29–19 ka), stagnant and downwasting ice remnants filled the main valleys during the earliest Lateglacial. This period, which formerly comprised the Bühl and Steinach Lateglacial stadials, is now defined as the phase of early Lateglacial ice decay (19-18 ka, in some sectors until 17 ka) [1]. Field evidence does not support the previously accepted view of a successive retreat of still nourished valley glacier tongues back into the Alps. Instead, the former Alpine-wide transection glacier complex collapsed rapidly as large parts transformed from former glacial accumulation to then ablation areas. The abundant meltwater in ice marginal and subglacial lakes led to an acceleration of ice loss.

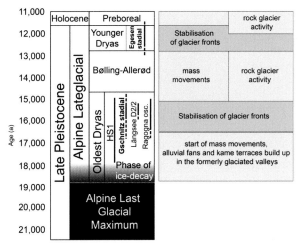

Fig. 1: *Stratigraphic scheme of the Alpine Lateglacial with the dominant processes for each time interval [1].*

During the Alpine Lateglacial (19–11.7 ka), two major glacier re-advances took place; the Gschnitz stadial and the Egesen stadial (Fig. 1). The first readvance of glaciers into a forefield free of dead-ice occurred during the Gschnitz stadial. The corresponding type locality is located at Trins in the Gschnitz Valley (Austria) where the

frontal moraine was ^{10}Be exposure dated to 16.8 ± 1.7 ka.

Fig. 2: *Right lateral Egesen stadial moraine complex of Rossboden glacier (Switzerland).*

The Younger Dryas Stadial was characterized by a sharp cooling after the relatively warm Bølling-Allerød Interstadial (Fig. 1). Alpine glaciers responded with significant advances in the tributary valleys where sets of stacked frontal moraines were constructed during the Egesen stadial (Fig. 2). Cosmogenic nuclide exposure dates suggest advances began already at the end of the Allerød period. Glacier stabilization and moraine build-up continued into the early Holocene. Maximum extents were attained early on, followed by reaching of successively smaller extents as the Younger Dryas period progressed. In some regions, rock glacier activity dominated, especially in the late Younger Dryas.

[1]   S. Ivy-Ochs et al., European Glacial Land-scapes: The Last Deglaciation (2023) 175

[1]CNR-IGG Padova, Italy
[2]Austrian Geological Survey, Vienna, Austria

POSTGLACIAL DENUDATION IN THE DORA BALTEA VALLEY

^{10}Be-derived catchment-wide denudation rates

E. Serra[1], P. G. Valla[2], R. Delunel[3], N. Gribenski[1], M. Christl, N. Akçar[1]

^{10}Be-derived catchment-wide denudation rates have been widely used to investigate the controlling mechanisms on recent (10^2-10^5 years) denudation dynamics [1]. ^{10}Be concentrations are measured in modern river sediments at the outlet of the studied basin and are inversely correlated with mean catchment denudation [2].

With the aim of further exploring the combined impacts of climatically-driven topography, tectonic uplift and bedrock erodibility on the efficiency of denudation processes, our study [3] investigated the spatial variability of ^{10}Be-derived denudation rates within the Dora Baltea (DB) catchment (western Italian Alps; Fig. 1).

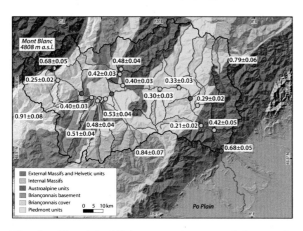

Fig. 1: *Simplified lithotectonic map of the study area with catchment-wide denudation rates (mm yr^{-1}) reported at sampling locations [3].*

We measured ^{10}Be concentrations on 18 river-sand samples collected along the DB river from the Mont Blanc Massif to the Po Plain and at the outlet of the main tributaries. ^{10}Be concentrations vary between 4.85±0.21 and 1.08±0.07 ×10^4 at g^{-1}, with inferred catchment-wide denudation rates between 0.2 and 0.9 mm yr^{-1}.

By statistically comparing denudation rates with topographic, environmental and geological metrics, we exclude a correlation of catchment denudation with modern precipitation and rock geodetic uplift. We found instead that catchment topography, in turn conditioned by bedrock structures and erodibility and glacial overprint, is the main driver of the spatial variability of denudation.

The lithotectonic and glacial controls on catchment denudation are exemplified by the Mont Blanc Massif. With the highest denudation rate (Fig. 1), it dominates the Dora Baltea sediment flux, explaining the constant low ^{10}Be concentrations measured along the DB course even downstream from the multiple junctions with tributary catchments (Fig. 2).

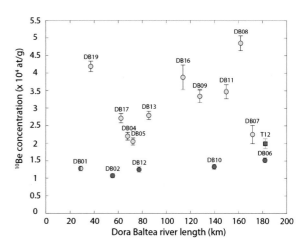

Fig. 2: *Downstream evolution of river-sand ^{10}Be concentrations in the DB catchment (red: samples collected along the main DB river; light blue: samples at the outlet of tributaries).*

[1] D. E. Granger and M. Schaller, Elements 10 (2014) 369

[2] F. von Blanckenburg, Earth Planet. Sci. Lett. 237 (2005) 462

[3] E. Serra et al., Earth Surf. Dyn. 10 (2022) 493

^{1}Geology, Univ. of Bern
^{2}ISTerre, Univ. of Grenoble Alps, France
^{3}EVS, Univ. Lumière Lyon, France

QUANTIFYING POST-GLACIER BEDROCK EROSION RATES

Erosion rates calculated using ^{10}Be and OSL in the European Alps

J. Elkadi[1], B. Lehmann[2], G.E. King[1], O. Steinemann, S. Ivy-Ochs, M. Christl, F. Herman[1]

The retreat of glaciers leaves an imprint on topography through glacial and non-glacial erosional processes. However, their specific mechanisms and relative impacts on the topography remains unclear. Currently in alpine environments, extensive research has investigated glacial erosion but fewer studies have examined non-glacial erosion. We applied a recently developed approach [1] which combines ^{10}Be with optically stimulated luminescence (OSL) to investigate bedrock post-glacier erosion rates adjacent to the Gorner Glacier, Switzerland (Fig. 1).

Fig. 1: *Study area and sampling sites (blue circles) highlighting the difference in surface preservation with elevation and thus exposure age.*

The results reveal erosion rates of the order of 10^{-2} to 10^{-1} mm a^{-1} [2]. These are in general agreement with bedrock erosion rates reported from studies with similar climates [e.g. 3]. We also observed a strong negative correlation between erosion rate and elevation but no correlation between erosion rate and slope [2] (Fig. 2). This suggests that frost crack weathering is perhaps not a dominant form of post-glacier weathering in this area. Similar trends in erosion rate with elevation and slope were reported at a nearby study from the Mont Blanc Massif [3] although the decrease in erosion rate with

elevation is more pronounced there. We suspect this difference is likely due to local variations, such as lithology and/or elevation, influencing the dominant post-glacier erosion mechanisms or the reflection of a potential relationship between erosion rate and exposure time.

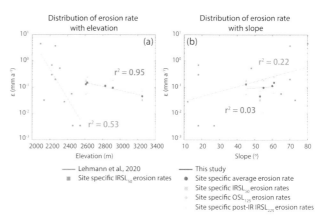

Fig. 2: *Distribution of bedrock surface erosion rates($\dot{\varepsilon}$) with elevation (a) and surface slope (b).*

Finally, we created a global compilation of both subglacial and periglacial erosion rates [2] (Fig. 2). Surprisingly, this revealed comparable orders of magnitude between the two processes although subglacial erosion rates remain higher. This could have important implications for landscape evolution models.

[1]   B. Lehmann et al., Earth Surf. Dyn. 7 (2019) 7

[2]   J. Elkadi et al., Earth Surf. Dyn. 10 (2022) 909

[3]   B. Lehmann et al., Geology 48 (2020) 139

[1]*Earth Surface Dynamics, Univ. Lausanne*
[2]*INSTAAR and Geology, Univ. Colorado, Boulder, USA*

USING P-PINI TO DECODE THE DECKENSCHOTTER

Dating the oldest Swiss Quaternary sediments with ^{10}Be and ^{26}Al

E. Broś, F. Kober[1], S. Ivy-Ochs, R. Grischott[2], M. Christl, C. Vockenhuber, P. Gautschi, C. Maden[3], J. D. Jansen[4], L. Ylä-Mella[4], M. F. Knudsen[5], J. Nørgaard[5], H.-A. Synal

The dynamic environment of high mountains with extensive and numerous glaciations, as well as repeated alternating phases of deposition and incision have created a complicated landscape on the northern Alpine foreland – the region where the oldest Quaternary glaciofluvial sediments in Switzerland (the Deckenschotter) are located.

To date the Swiss Deckenschotter, we have measured ^{10}Be and ^{26}Al in 77 samples from seven different gravel units at six outcrops across the Swiss foreland. For the age calculations, we implement a new burial-dating model called P-PINI (Particle Pathway Inversion of Nuclide Inventories) [1, 2]. The P-PINI method takes into account i) non-steady erosion in the source area, ii) discontinuous exposure prior to burial and iii) the altitude dependence of ^{26}Al/^{10}Be pre-burial (surface) ratio. The model applies a source-to-sink approach creating a library of possible virtual samples based on the input parameters defining the source and the sink. The input parameters are randomized in Monte Carlo simulations. The created library is further compared with the AMS-measured concentrations of ^{10}Be and ^{26}Al in the samples (Fig 1). The P-PINI code evaluates the probability distribution of the simulated versus the measured samples and accepts only plausible scenarios for each data point. This rejection sampling yields the most probable age for the dated Deckenschotter horizon.

The results of modelling with the P-PINI code will help further examination and refinement of the question of the age of the various Deckenschotter units in the northern Swiss Alpine foreland. This region is designated for a geological disposal facility of radioactive waste. Thus, the time of Deckenschotter deposition is important for understanding long-term landscape evolution and for ensuring safe long-term storage of such nuclear radioactive waste.

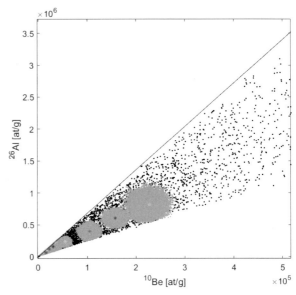

Fig. 1: ^{26}Al-^{10}Be library with millions of simulated samples (black dots) compared with eleven AMS-measured samples (coloured stars) from the site Hungerbol, resulting in an inventory of accepted simulations (orange dots). Black line shows the ^{26}Al/^{10}Be ratio of 6.8.

[1] M. F. Knudsen et al., Earth Planet. Sci. Lett. 549 (2020) 116491

[2] J. Nørgaard et al., Quat. Geochronol. 74 (2023) 101420

[1]NAGRA, Wettingen
[2]BTG AG, Büro für Technische Geologie, Sargans
[3]Geochemistry and Petrology, ETH Zurich
[4]Geophysics, Czech Acad. of Sci., Prague, Czech Republic
[5]Geoscience, Aarhus Univ., Denmark

EARLY PLEISTOCENE STRATIGRAPHY IN THE ALPS

Reconstructing the Deckenschotter chronology with ^{10}Be and ^{26}Al

C. Dieleman[1], M. Christl, C. Vockenhuber, P. Gautschi, N. Akçar[1]

The Swiss northern Alpine foreland was cannibalized by 15 glacier advances during the Quaternary [1]. The Deckenschotter represent the oldest Quaternary deposits in the foreland and are characterized by a succession of glaciofluvial sediments (Fig. 1). Their chronostratigraphy remained relatively unconstrained for a long time until the application of cosmogenic nuclides to shed light on their timing. Recent studies suggest ages of ca. 2 Ma and ca. 1 Ma [e.g. 2, 3]. In this study, we focused on four new Deckenschotter sites at Irchel and in the Lake Constance region for better a understanding of how the landscape evolved over the past 2.6 Ma.

Fig. 1: *Photograph of the Schartenflue outcrop at Irchel.*

At the studied Deckenschotter outcrops the sediments were analyzed in detail to reconstruct the provenance of the gravels, mechanisms of their transport, paleoflow directions, and their depositional environments. Isochron-burial dating with cosmogenic ^{10}Be and ^{26}Al was applied to establish the chronology.

The clast petrographical composition indicates a provenance from the Central and eastern Central Alps. Most of these gravels show a predominant glacial transport and only few a fluvial one (Fig. 2).

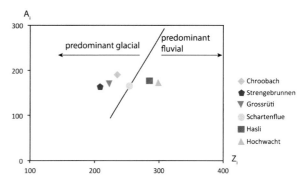

Fig. 2: *Clast morphometry results [4].*

Our isochron-burial ages in comparison with the already existing chronology reveal that the Deckenschotter were deposited within three major gravel accumulation phases: the first one at ca. 2.5 Ma, the second one around 1.5 Ma and the third one around 1 Ma [4]. In addition, these ages suggest a more complex cut-and-fill architecture at Irchel [4].

[1] C. Schlüchter, Bulletin de l'Association Française pour l'étude du Quarternaire, 25 (1988) 141

[2] N. Akçar et al., Earth Surf. Process. Landf. 42 (2017) 2414

[3] A. Claude et al., Geol. Soc. Am. Bull. 131 (2019) 2056

[4] C. Dieleman et al., Swiss J. Geosci. 115 (2022) 11

[1]Geology, Univ. Bern

RECONSTRUCTING THE TAMINS ROCK AVALANCHE

^{36}Cl boulder exposure dating and runout modelling

O.A. Pfiffner[1], S. Ivy-Ochs, Z. Mussina, J. Aaron[2], O. Steinemann, C. Vockenhuber, N. Akçar[1]

Combining detailed field survey, sediment and landform analysis, cosmogenic nuclide surface exposure dating, and runout modeling we reconstructed the timing and dynamics of the Tamins rock avalanche [1]. The Tamins rock avalanche deposits are located just a few kilometers downstream of the Flims rock avalanche deposits.

Fig. 1: *View of the horseshoe-shaped niche just below the peak Säsagit.*

The morphology of the deposits (volume 1.2 km^3) suggests that the Tamins rock avalanche broke away as a semicoherent block, which subsequently sheared due to a slower moving base. Spreading of the rock avalanche body resulted in extensional scarps that separate large (1–2 km long) W-E trending transverse ridges (Rascheu, Carschitscha) (Fig. 1). At the Fanaus outcrop (see photo at the beginning of the Cosmogenic Nuclides section) meter-sized blocks of the carapace overlie finely crushed limestone with smaller blocks of the main body of the deposit and witness the intense fragmentation.

Modelling results suggest that the Tamins rock avalanche likely travelled for a significant distance as a flexible block, before fragmenting and turning flowlike. Our simulations show that the mass impacted the valley floor after about 40 s, and the mass spread both to the southeast

and southwest following fluidization. Thick deposits at Rascheu are apparent in the simulation (Fig. 2), which reasonably reproduce field observations.

Exposure dating of eight boulders located all across the Tamins rock avalanche deposits yields an age of the Tamins rock avalanche of 9420 ± 880 years BP. The age of the Tamins event is close to but slightly older than the age of the neighboring Flims rock avalanche (9475–9343 cal BP [2]). The ages, as well as the morphology and internal structure of the Tamins rock avalanche, suggest that it occurred as a single event.

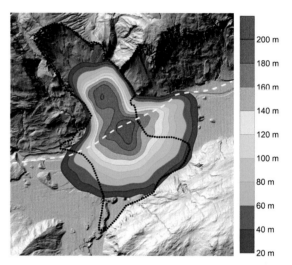

Fig. 2: *Best-fit deposit extent and depths obtained with Dan3D-Flex. Dashed black outline shows the post-event deposit area; dashed white line shows the location of the change from the frictional rheology to the Voellmy rheology used in the model.*

[1]   O. A. Pfiffner et al., Landslides (2022)
[2]   K. Nicolussi et al., Geomorphology 243 (2015) 87

[1]*Geology, Univ. Bern*
[2]*Earth Sciences, ETH Zurich*

LGM GLACIAL ADVANCES IN BARHAL VALEY (NE, TÜRKIYE)

Surface exposure dating of boulders with ^{36}Cl

R. Reber[1], N. Akçar[1], D. Tikhomirov[1,2], S. Yesilyur[1,3] C. Vockenhuber, V. Yavuz[4], S. Ivy-Ochs, C. Schlüchter[1]

Barhal Valley belongs to the Çoruh Valley System in the Kaçkar Mountains of north-eastern Anatolia. Todays moisture source of an average yearly precipitation of 2000 mm is the Black Sea, situated approximately 40 km to the north of the study site. Glaciers of the Last Glacial Maximum (LGM) descended directly from Mt. Kaçkar (3932 m a.s.l.) and reached an altitude of ca. 1850 m a.s.l. The position of Barhal Valley to the south of the main weather divide and its east–west orientation has a strong influence on the existence and expansion of paleoglaciers.

Fig. 1. *Glacier reconstruction in the Central Çoruh Valley System for the 22.2 ± 2.6 ka advance [1].*

To this conclusion we evolved after obtaining 32 new cosmogenic ^{36}Cl dates on erratic boulders.

With the age constrain on the boulders and the geomorphological mapping of the area, three glacier advances could be reconstructed in the Barhal Valley, namely at 34.0 ± 2.3 ka, 22.2 ± 2.6 ka, and 18.3 ± 1.7 ka within the time window of the global LGM. Field evidence shows that the glacier of the 18.3 ± 1.7 ka advance disappeared rapidly and that by the latest time, at 15.6 ± 1.8 ka, the upper cirques were ice-free. No evidence for Lateglacial glacier fluctuations was found, and the Neoglacial activity is restricted to the cirques with rock glaciers. A range of 2700 to 3000 m for the Equilibrium Line Altitude (ELA) at the LGM was reported based on modelling of the glacial morphology (Fig. 1). We determined that the most likely position of the LGM ELA in the Barhal Valley was at 2900 m a.s.l.

We suggest an alternative moisture source to the direct transport from the Black Sea for the ice accumulation in the Eastern Black See Mountains. The shift of the Polar Front and of the Siberian High Pressure System to the south during the LGM resulted in the domination of easterly airflow to the Caucasus and Kaçkar Mountains with moisture from expanded lakes in central–western Siberia and from the enlarged Aral- and Caspian Seas.

[1]   R. Reber et al., Geosciences 12 (2022) 257

[1]*Geology, Univ. Bern*
[2]*Geography, Univ. Zurich*
[3]*Geography, Ankara Univ., Türkiye*
[4]*Civil Engineering, Turkish-German Univ. Istanbul, Türkiye*

^{10}Be IN BLACK SEA SEDIMENTS DURING TERMINATION II

A synchronization tool for multi-archive paleoclimate studies?

M. Czymzik[1], O. Dellwig[1], H.W. Arz[1], R. Muscheler[2], M. Christl

A ^{10}Be record from Black Sea sediments (Fig. 1) provided a well-preserved time-series of cosmogenic radionuclide production rate changes associated with the Laschamp geomagnetic excursion [1]. This time-series allowed to synchronize the sedimentary ^{10}Be record to that from Central Greenland ice cores using curve fitting and investigate time-transgressive climate changes across the Northern Hemisphere connected with Greenland Interstadial 10 [1].

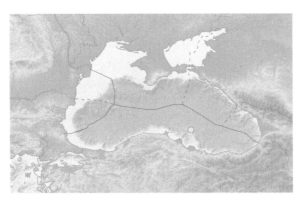

Fig. 1: *Bathymetric map of the Black Sea with location the investigated sediment core M72/5-22 GC8.*

To further explore the potential of ^{10}Be in this archive, we measured a ^{10}Be record from Black Sea sediment core M7/5-22 GC8 from 805 to 940 cm core depth covering the penultimate glacial/Eemian transition ~127-133.5 ka ago (Termination II) (Fig. 2). Possible environmental influences on ^{10}Be deposition during this period were assessed and reduced by calculating authigenic ^{10}Be/^{9}Be-ratios.

The resulting ^{10}Be and ^{9}Be records reveal cm-scale fluctuations (Fig. 2). A negative ^{10}Be and ^{9}Be excursion from 850 to 859 cm core depth flanked by distinct positive peaks occurs around the position of the P-11 tephra [2]. The ^{10}Be/^{9}Be record indicates cm-scale variability from 945 to 850 cm core depth, accompanied by an increasing trend from 850 to 805 cm core depth.

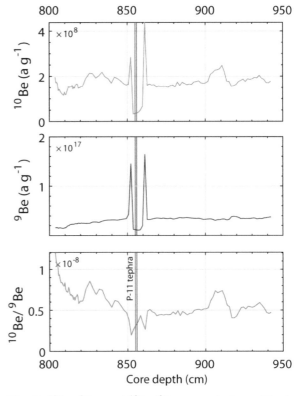

Fig. 2: *^{10}Be, ^{9}Be and ^{10}Be/^{9}Be records from Black Sea sediment core M72/5-22 GC8 covering Termination II. The position of the P-11 tephra is indicated.*

Future investigations of the measured ^{10}Be record from Black Sea sediments around Termination II will target on its potential as synchronization tool and for multi-archive climate studies.

[1] M. Czymzik et al., Proc. Natl. Acad. Sci. U.S.A. 117 (2020) 28649

[2] A. Wegwerth et al., Earth Planet. Sci. Lett. 404 (2014) 135

[1]Marine Geology, Baltic Sea Research Warnemünde, Germany
[2]Geology, Lund Univ., Sweden

INTERNAL DEFORMATION OF THE ANATOLIAN SCHOLLE

Dating offset fluvial deposits across the Ovacık Fault, Türkiye

C. Zabcı[1], T. Sançar[2], D. Tikhomirov[3,4], S. Ivy-Ochs, C. Vockenhuber, A.M. Friedrich[5], M. Yazıcı[1,6], N. Akçar[3]

The active tectonics of Anatolian Scholle is mostly characterized by its westward motion with respect to Eurasia. Although most of the deformation is suggested to be confined along the Anatolia's boundary elements (the North and East Anatolian shear zones) recent studies indicate a higher magnitude of internal strain accumulation, especially along the parallel/sub-parallel strike-slip faults of its central province.

We present the first morphochronology-based slip rate estimate for one of these strike-slip structures, the Ovacık Fault by using cosmogenic ^{36}Cl dating of offset fluvial deposits at the Köseler site (39.3643°N, 39.1688°E) (Figs. 1 and 2) [1].

Fig. 1: *Photograph showing the alluvial fan surface and the inset terraces of the Köseler Site.*

A faulted riser, bounding the alluvial fan (NF1) and the inset terrace tread (T2), is displaced 15-22 m. The scattered surface ages and variability of ^{36}Cl concentrations in depth profiles suggest strong evidence for inheritance in alluvial fan and terrace deposits, thus we used modelled depth-profile ages for both surfaces. The modelled ages of 6-8 ka for T2 yield a slip rate estimate of 2.8 + 0.7/- 0.7 mm a^{-1} for the lower-tread reconstruction of the NF1/T2.

Our results together with previous slip rate estimates for other structures show significant internal deformation for Anatolia, especially along its sub-parallel strike-slip faults [1]. These secondary faults slice Anatolia into several pieces giving rise to the formation of the Malatya-Erzincan, Cappadocian and Central Anatolian slices, where the geometry is strongly controlled by the distribution of the Tethyan accretionary complexes.

Fig. 2: *(a) Partly buried cobbles and boulders which are densely distributed along both alluvial fan and terrace surfaces. (b) Trench in the T2 terrace for depth-profile sampling.*

[1]   C. Zabci et al., Tur. J. of Earth Sci. (2023) in press

[1]*Jeoloji, Istanbul Teknik Üniversitesi, Türkiye*
[2]*Coğrafya, Munzur Üniversitesi, Türkiye*
[3]*Geology, Univ. Bern*
[4]*Geography, Univ. Zurich*
[5]*Geo- und Umweltwissenschaften, LMU, München, Germany*
[6]*GFZ, Potsdam, Germany*

CONTROLS ON RIVER INCISION AND AGGRADATION

^{36}Cl and ^{10}Be paleo-denudation rates on Crete

R.F. Ott[1], D. Scherler[1,2], K.W. Wegmann[3], M.K. D'Arcy[4], R.J. Pope[5], S. Ivy-Ochs, M. Christl, C. Vockenhuber, T.M. Rittenour[6]

Rivers adjust their slope and river bed based on current water and sediment discharge conditions. If water discharge decreases, or the sediment discharge increases, rivers tend to aggrade and form alluvial plains. When water discharge increases and/or sediment discharge is lower, rivers will respond by incision. Climate change is thought to affect both, water discharge patterns by changing precipitation and sediment discharge, by changing the rate of denudation. However, it is unknown whether alluvial sedimentary archives mostly record variations in water or sediment discharge.

We used optically stimulated luminescence (OSL) to date aggradation and incision periods in two alluvial fan sequences in southwestern Crete during the last glacial cycle [1]. Additionally, we measured concurrent paleo-denudation rates with ^{10}Be and ^{36}Cl measurements, to link aggradation-incision dynamics to sediment discharge (Fig. 1).

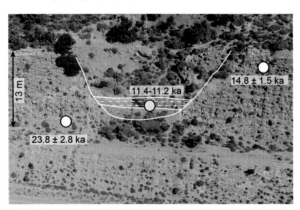

Fig. 1: *One of the dated alluvial aggradation and incision sequences from western Crete.*

The studied alluvial sequences show aggradation during interstadials of marine isotope stages (MIS) 2, 4, and possibly 6, and incision during interglacials and interstadials of MIS 1, 3, and 5e (Fig. 2). At the same time, the paleo-denudation

rates show only minor variations. This suggests that despite major shifts between aggradation and incision, the supply of sediment from the hillslopes remained relatively steady. Therefore, we conclude that on Crete, variations in water discharge outcompeted changes in sediment discharge and are the primary driver of the aggradation-incision behavior of rivers.

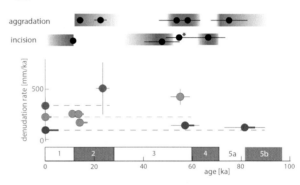

Fig. 2: *Periods of aggradation and incision (upper panel) compared against measured paleo-denudation rates.*

[1] R. F. Ott et al., Earth Surf. Process. Landf. (2022) 1

[1]*GFZ, Potsdam, Germany*
[2]*Geological Sciences, Freie Universität Berlin, Germany*
[3]*Marine, Earth and Atmospheric Sciences & Center for Geospatial Analytics, North Carolina State Univ., Raleigh, USA*
[4]*Earth, Ocean and Atmospheric Sciences, University of British Colombia, Vancouver, Canada*
[5]*School of Environmental Sciences, College of Built and Natural Environment, Univ. Derby, UK*
[6]*Geosciences, Utah State Univ., Logan, USA*

SYNCHRONOUS RECORDS OF PAST BALTIC CHANGE

[10]Be in Baltic Sea and Lake Kälksjön sediments around 5500 a BP

M. Czymzik[1], O. Dellwig[1], H.W. Arz[1], R. Muscheler[2], M. Christl

Aligning common variations in the cosmogenic radionuclide production rate provides a tool for the synchronization of natural environmental archives. Using this approach, we aim at synchronizing new [10]Be records from Western Gotland Basin (WGB, Baltic Sea) and Lake Kälksjön (KKJ, central Sweden) sediments to the ^{14}C production time-series derived from the IntCal20 calibration curve [1, 2] during the Mid-Holocene time-interval ~6400 to 5200 a BP using semi-automatic curve fitting [3].

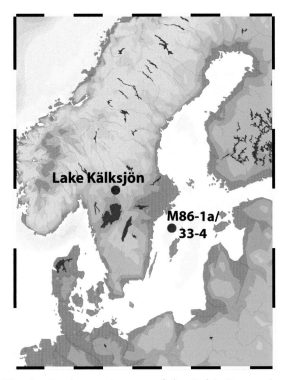

Fig. 1: *Bathymetric map of the Baltic Sea region with locations of the investigated sediment cores from Lake Kälksjön and Western Gotland Basin.*

The synchronizations to the IntCal20 ^{14}C production time-scale point to decadal to multi-decadal adjustments of the WGB (not shown) and KKJ chronologies (Fig. 2). They most likely place both sediment records into the context of the IntCal20 ^{14}C time-scale and reduce the originally centennial-scale chronological uncertainties to about ±20 yr (WGB) and ±40 yr (KKJ).

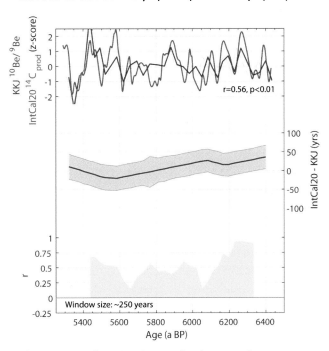

Fig. 2: *Synchronization of the environment corrected [10]Be record ([10]Be/[9]Be-ratio) from Lake Kälksjön sediments to the ^{14}C production time-series from the IntCal20 calibration curve based on semi-automatic curve-fitting.*

Such placement on one time-scale with reduced chronological uncertainties opens up the possibility for investigating possible time-transgressive environmental changes, with high temporal precision.

[1]   R. Muscheler et al., Quat. Sci. Rev. 24 (2005) 1849
[2]   P. Reimer et al., Radiocarbon 62 (2020) 725
[3]   M. Czymzik et al., Proc. Natl. Acad. Sci. U.S.A. 117 (2020) 28649

[1]*Marine Geology, Baltic Sea Research Warnemünde, Germany*
[2]*Geology, Lund Univ., Sweden*

DRILL CHIPS FROM LITTLE DOME C AND SOLAR ACTIVITY

Using a Bayesian method to separate the solar and geomagnetic signal

L. Nguyen[1], A. Nilsson[1], M. Christl, P. Gautschi, R. Mulvaney[2], J. Rix[2], R. Muscheler[1]

Cosmogenic radionuclides provide the best proxy records for solar activity reconstructions over centennial to millennial time scales. However, such reconstructions can suffer from an incomplete understanding of the carbon cycle influence on ^{14}C or atmospheric circulation and deposition processes on the deposition of ^{10}Be in polar ice cores. Differences between these records are still an unresolved issue for long-term solar activity reconstructions [e.g. 1]. Furthermore, the correction of geomagnetic field influences on the radionuclide records presently still leads to large uncertainties.

This project aims to improve Holocene solar activity reconstructions by (i) obtaining a high-quality and continuous ^{10}Be record from East Antarctica where weather and climate influences on the ^{10}Be deposition can be assumed to be small and (ii) by applying a novel Bayesian method [2] to separate the solar from the geomagnetic field signal.

Fig. 1 shows the last 2000 years of ^{10}Be data from Little Dome C (LDC), the place where presently the Beyond-EPICA deep ice core drilling is ongoing (panel c). The data is measured on drill chips [3] from an exploratory drilling to find a suitable site for obtaining an ice core reaching 1.5 million years back in time. Panel a shows the solar signal, panel b the corresponding geomagnetic dipole field intensity and panel c the LDC ^{10}Be data. The solar and geomagnetic signals were separated using a Bayesian model [3] that includes prior knowledge on the typical time scales on which the geomagnetic dipole field and solar shielding vary. Differences to independent geomagnetic field reconstructions [4] (Fig 1b) indicate possible normalization issues, incorrect assumptions in the prior for the Bayesian model, uncorrected transport/climate impacts on ^{10}Be or point to uncertainties in the geomagnetic field models.

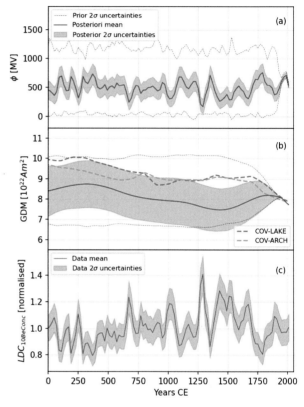

Fig. 1: *Solar (a) and geomagnetic field (b) reconstructions inferred from ^{10}Be measurements on drill chips from little Dome C (c).*

In this project we will expand this analysis over the complete Holocene period in an effort to provide the most robust solar activity reconstruction to date.

[1] M. Vonmoos et al., J. Geophys. Res. Space Phys. 111 (2006) A10105

[2] L. Nguyen et al., Earth Planets Space 74 (2022) 130

[3] L. Nguyen et al., Results in Geochemistry 5 (2021) 100012

[4] G. Hellio & N. Gillet, Geophys. J. Int. 214 (2018) 1585-1607

---

[1]*Quaternary Sciences, Lund Univ., Sweden*
[2]*British Antarctic Survey, Cambridge, UK*

THE ^{10}Be/^{36}Cl RATIO AS A DATING TOOL

Towards dating the oldest ice core with cosmogenic radionuclides

N. Kappelt[1], R. Muscheler[1], M. Baroni[2], G. Raisbeck[3], F. Yiou[3], C. Vockenhuber, M. Christl

Scientists of the "Beyond EPICA: Oldest Ice Core" (BE-OIC) project are currently drilling an ice core at Little Dome C in Antarctica (LDC, 75.30°S, 122.45°E) in an effort to retrieve a continuous, 1.5-million-year-old ice core record (Fig. 1). Correctly dating the core is essential for interpreting any obtained data. Because many dating methods rely on assumptions about climate synchronicity and require the correct identification and synchronisation of signals, such methods are complicated for the bottom part of the core, where extreme layer thinning occurs, and the continuity may be disturbed.

One possible tool for obtaining an independent age estimate is the ^{36}Cl/^{10}Be ratio, which has an effective half-life of 384'000 years. The radionuclides ^{10}Be and ^{36}Cl are produced by interactions of cosmic rays with molecules in the atmosphere and, by measuring their ratio, the production variability should be largely removed [1]. What can alter the ratio is ^{36}Cl loss due to outgassing at low accumulation sites, such as EPICA Dome C (EDC) and LDC under present day conditions [2]. However, preliminary measurements have shown that limited to no loss occurs during the last glacial maximum, when dust concentrations were high and neutralised the acidic species responsible for the outgassing, a mechanism which also preserves sea-salt chlorine [3].

In this project we aim to test whether ^{36}Cl was also preserved in previous glacial periods and if the decay curve of the ^{10}Be/^{36}Cl ratio aligns well with the expected signal based on the current timescale of the EDC core.

For this purpose, we analysed ice and meltwater samples from the EDC core with ages ranging from the Holocene back to 746'000 years BP, most of which lie within periods of favourable conditions for chlorine preservation: high dust concentrations as estimated by the non-sea-salt calcium (nss-Ca2+).

Fig. 1*: Ice core segment from Little Dome C. Photo credit: Robert Mulvaney from British Antarctic Survey.*

concentration and a stable Cl$^-$/Na$^+$ ratio close to the sea-salt reference value. Additional samples were measured from periods with expected chlorine loss to test whether low nss-Ca^{2+} concentrations and Cl$^-$/Na$^+$ ratios also indicate chlorine loss in older samples. Additionally, ice shaving and meltwater samples from similar depths were taken for an intercomparison of the two sample types.

While a decay signal is expected to be evident in the measured samples, it may be disturbed due to varying climate conditions, which could affect the ^{10}Be and ^{36}Cl transport and deposition in different ways.

[1]   G. Wagner et al., NIM-B 172 (2000) 597
[2]   R. J. Delmas et al., Tellus 56B (2004) 492
[3]   R. Röthlisberger et al., J. Geophys. Res. Atmos. 108 (2003) 4526

[1]Quaternary Sciences, Lund Univ., Sweden
[2]CEREGE, Aix-Marseille Univ., France
[3]IJCLab, CNRS, Univ. Paris-Saclay, France

THE SOLAR SIGNAL IN ICE CORE EXCESS WATER SAMPLES

Comparing the solar cycle in traditional and novel ice core samples

C. Paleari[1], F. Mekhaldi[1], T. Erhardt[2], M. Zheng[1,3], M. Christl, F. Adolphi[2], M. Hörhold[2], R. Muscheler[1]

Recent results showed that [10]Be measurements in excess water from continuous flow analysis (CFA) can be used to identify solar storm signals in ice cores [1]. However, the method of using these samples, usually considered as waste water, has never been systematically assessed. Here we did parallel [10]Be measurements in clean ice and in CFA excess water samples in the S6 firn core from the EGRIP ice core project [2].

The CFA excess water samples show a very similar signal as the clean ice/firn samples supporting the viability of this sampling method. It enables us to obtain high-resolution [10]Be records with a relatively efficient sampling method without the need for competing for the valuable ice from deep ice core drilling projects [2].

In addition, we investigate the solar 11-yr cycle in the S6 data. As the solar cycle is actually a 22-yr magnetic cycle, the polarity of the solar magnetic field changes every 11 years. In contrast to the sunspot number, the cosmic ray modulation shows very different expressions depending on the polarity of the sun (as the polarity determines the preferred pathways of galactic cosmic rays within the heliosphere). An example for the results is shown in Fig. 1 [2].

We can conclude from this analysis that it is possible to identify the shape of the solar cycle in discrete high-resolution [10]Be samples as well as in CFA excess water samples. However, the uncertainties for such investigations should not be underestimated. For example, the analysis for the even cycles (which show sharper peaks in the cosmic ray flux to Earth) shows larger disagreements between theoretical expectation and the data from the EGRIP S6 core [2].

Similarly, our study shows that we cannot robustly see the relatively small signal induced by known solar storms over the past 70 years in

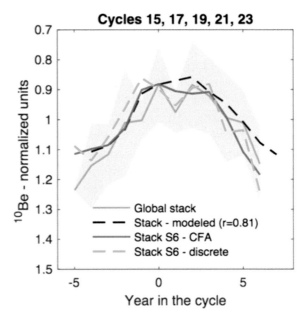

Fig. 1: *Cosmic ray intensity as seen by neutron monitors for odd solar cycles 15, 17, 19, 21, and 23 (dashed black line, superimposed epoch analy–sis). The relatively broad peak is also visible in a global stack of [10]Be records (orange). Considering the uncertainties the expression in the CFA (pink) and discrete (green) samples is very similar [2].*

the high-resolution radionuclide data. Progress in the understanding of confounding factors (e.g. weather, accumulation rate influences or possible volcanic influences on the [10]Be deposition) are needed to be able to better understand the relatively large scatter in the [10]Be deposition over seasonal time scales [3].

[1] C. Paleari et al., Nat. Comm. 13 (2022) 214
[2] C. Paleari et al., Clim. Past Disc. in review
[3] M. Zheng et al., EPSL 541 (2020) 116273

[1]*Quaternary Sciences, Lund Univ., Sweden*
[2]*Alfred Wegener Institute, Bremerhaven, Germany*
[3]*Earth Systems Sciences, ETH Zurich*

ANTHROPOGENIC RADIONUCLIDES

Sunrise in the Canada Basin of the Arctic Ocean, JOIS expedition, September 2022

^{14}C in the Western Mediterranean Sea in 2022

^{129}I and ^{236}U in the Norwegian Coastal Current

^{129}I from the Arctic Century Expedition 2021

Results from the Arctic Ventilation Cruise 2021

Preliminary results from JOIS 2020 & 2021

2022 JOIS expedition

Tracing ocean circulation around Iceland

^{129}I in the Gulf of Saint Lawrence (Canada)

2022 expeditions to Labrador Sea and Davis Strait

^{129}I and ^{236}U from BOCATS and AR7W expeditions

^{129}I and ^{236}U in the subtropical North Atlantic

First ^{129}I measurement on Milea

Anthropogenic U in the Southern Atlantic Ocean

Reactor-derived vs environmental plutonium

Combining AMS and DGT for actinide measures

Fluoride target matrices for ^{239}Pu analyses

Single-sample analysis of actinide isotopes

^{14}C IN THE WESTERN MEDITERRANEAN SEA IN 2022

Preliminary results of radiocarbon's temporal variability in the basin

L. Raimondi[1], L. Wacker, N. Casacuberta[1]

Just like an ocean, the Mediterranean Sea (MedSea) is characterized by regions of deep-water formation. Nevertheless, this process occurs at a much faster rate here than in the global ocean. This peculiarity makes the MedSea a perfect laboratory to study ocean circulation and predict future changes. The TAlPro 2022 cruise, that took place in May 2022 on board of the *RV Belgica*, aimed at re-occupying stations in the Western MedSea to monitor the temporal variability of the MedSea circulation.

During this cruise we collected samples for the measurement of three different isotopes (^{129}I, ^{236}U and ^{14}C). Here we only show preliminary results of the ^{14}C, with particular focus on the distribution of ^{14}C along a north-to-south transect which extends from the Ligurian Sea to the coast of Sicily and further we provide a first insight into the temporal variability of ^{14}C between the 2011 [1] and 2022 oceanographic campaigns.

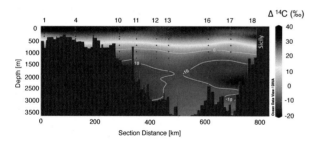

Fig. 1: *North-South transect from Ligurian Sea to Sicily of Δ^{14}C from the TAlPro 2022 Cruise.*

In 2022, Δ^{14}C ranged between -18 and 38‰ in the Western Mediterranean, a narrow range if compared to the overall ^{14}C bomb peak. The highest values were found in the upper 500 m of the water column, while a minimum was located around the 1500-2000 m depth range, followed by a slight increase near the bottom (Fig. 1).

In Fig. 2 we show the vertical distribution of Δ^{14}C along two repeated stations. In both stations we observed a temporal increase of Δ^{14}C below 500 m which reflects the delayed bomb peak signal reaching deeper in the water column. On the other hand, a decrease in Δ^{14}C occurred in the surface layer (Tyrrhenian Sea station) which is the result of the combined effect of the vertical transport of the bomb-peak signal due to deep water convection and the dilution effect due to increased anthropogenic CO_2 emissions (i.e. the Suess Effect).

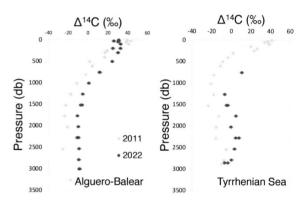

Fig. 2: *A) Map of the MedSea, with locations of ^{14}C stations (2011 in red and 2022 in yellow dots). B) vertical profiles of two repeated stations occupied both in 2011 and 2022.*

[1] T. Tanhua et al., Earth Syst. Sci. Data 5 (2013) 294

[1]*Environmental Systems Sciences, ETH Zurich*

^{129}I AND ^{236}U IN THE NORWEGIAN COASTAL CURRENT

Constraining input functions to the Arctic Ocean

A.-M. Wefing[1], N. Casacuberta[1], C. Vockenhuber, M. Christl, K. Kündig[1]

The anthropogenic radionuclides ^{129}I and ^{236}U are discharged by two European nuclear fuel reprocessing plants (RPs) in Sellafield (SF, UK) and La Hague (LH, France). Both have been proven to be powerful tracers of Atlantic Water in the Arctic Ocean and further downstream, reaching as far as the subtropical North Atlantic [1, 2]. Most of the RP discharges mix in the North Sea and flow northwards following the Norwegian Coastal Current (NCC).

Input functions for ^{129}I and ^{236}U are defined at the Barents Sea opening (north of Norway) and reflect the tracer concentrations entering the Arctic Ocean over time (Fig. 1) [3]. These input functions were based on a dataset from 2015, taking the time-varying discharges from the RPs and assuming a constant dilution factor.

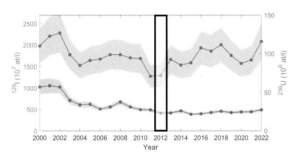

Fig. 1: *NCC input function defined at the Barents Sea opening for ^{129}I (blue) and ^{236}U (red).*

In this study, surface samples collected along the Norwegian Coast in 2012 were analyzed for ^{129}I and ^{236}U to better understand the northward transport of the tracer signal and to review the NCC input function. For both radionuclides, highest concentrations were found around 60°N, where the signal advected from the North Sea meets the Norwegian Coast (Fig. 2). The northward decrease in the signal was similar for both tracers, whereas differences were observed in the Baltic Sea. Here, the ^{236}U concentrations are much higher compared to ^{129}I, potentially due

to the higher input of ^{236}U from nuclear weapon testing in the mid-20th century as also suggested in [4].

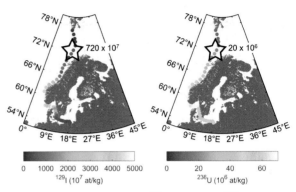

Fig. 2: *Surface maps of ^{129}I (left) and ^{236}U (right) concentrations along the NCC. The star marks the location where the input function (Fig. 1) is defined and the corresponding concentrations.*

The NCC input function (Fig. 1) was defined at about 72°N (Fig. 2, star) and yields ^{129}I and ^{236}U concentrations of about 1400×10^7 at L^{-1} and 25×10^6 at L^{-1}, respectively, for the year 2012. While the ^{236}U concentration was supported by the new dataset, the measured ^{129}I concentration in the NCC in 2012 was only about half (Fig. 2). This could be due to the sampling location slightly offshore, suggesting that ^{129}I is more confined to the coast due to the fact that it is mostly released by LH. On the contrary, ^{236}U has mainly been released by SF, and also has the input from weapon tests.

[1]   N. Casacuberta and J. N. Smith, Annu. Rev. Mar. Sci. 15 (2023) 203

[2]   A.-M. Wefing et al., LIP annual report (2022) 76

[3]   N. Casacuberta et al., J. Geophys. Res. Oceans 123 (2018) 6909

[4]   M. Lin et al., Water Res. 210 (2022) 117987

[1]Environmental Systems Sciences, ETH Zurich

^{129}I FROM THE ARCTIC CENTURY EXPEDITION 2021

A key study area for Atlantic branches entering the Arctic Ocean

A.-M. Wefing[1], N. Casacuberta[1], C. Vockenhuber, M. Christl, K. Kündig[1]

The Barents and Kara Seas are key study areas to investigate how different branches of Atlantic Water enter the Arctic Ocean (Fig. 1). The anthropogenic radionuclide ^{129}I, mainly released from European nuclear reprocessing plants, has proved to label Fram Strait Branch Water (FSBW), Barents Sea Branch Water (BSBW) and the Norwegian Coastal Current (NCC) (Fig. 1) with different concentrations at each branch. This makes ^{129}I an ideal tracer to study pathways and mixing of these Atlantic branches in the central Arctic Ocean [1].

In this study, samples for ^{129}I were collected during the Arctic Century Expedition in 2021, which was a joint effort between the Swiss Polar Institute (SPI), the Arctic and Antarctic Research Institute (AARI, Russia), and the Helmholtz Centre for Ocean Research Kiel (GEOMAR, Germany) (Fig. 1).

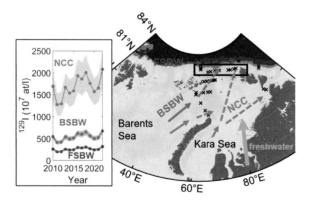

Fig. 1: *^{129}I input functions in the three Atlantic branches defined at the Barents Sea Opening and locations of stations sampled for ^{129}I in 2021. The black rectangle outlines the section plotted in Fig. 2.*

^{129}I concentrations along a section through the St. Anna Trough showed significant spatial variability that can be related to the presence of the different branches (Fig. 2). Highest ^{129}I concentrations of about 700 x 10^7 at L^{-1} were found at the eastern end of the trough, pointing to a high fraction of NCC waters which carry the strongest tracer signal (Fig. 2). However, these concentrations are lower than the pure NCC input function signal expected for 2021 (Fig. 1), which can be explained by the dilution of the tracer signal with freshwater, e.g., from the Ob river, or mixing with BSBW in the Barents Sea.

In the western part of the St. Anna Trough, ^{129}I concentrations were around 500 x 10^7 at L^{-1}, which fits well to the tracer signal expected in BSBW. The same holds true for stations within FSBW (not shown here) and confirms the ^{129}I input functions in both branches that were estimated for the Barents Sea Opening, i.e., upstream of the Kara Sea.

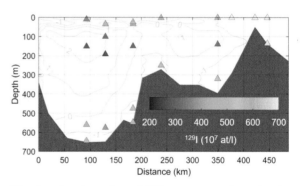

Fig. 2: *Section plot of ^{129}I concentrations in the St. Anna Trough (black rectangle in Fig. 1).*

For most stations, samples have also been collected for the analysis of ^{236}U, which is still ongoing. The combined dataset of both tracers, potentially coupled with other tracers such as stable oxygen or Neodymium isotopes, will allow to disentangle different sources of dilution in the study area and quantify mixing processes.

[1] N. Casacuberta et al., J. Geophys. Res. Oceans 123 (2018) 6909

---

[1]Environmental Systems Sciences, ETH Zurich

RESULTS FROM THE ARCTIC VENTILATION CRUISE 2021

An update on ^{129}I and ^{236}U in the central Arctic Ocean

A.-M. Wefing[1], N. Casacuberta[1], C. Vockenhuber, M. Christl, K. Kündig[1]

In 2021, an expedition to the Eurasian Basin of the Arctic Ocean was conducted as part of the Synoptic Arctic Survey (SAS) program (Fig. 1, black box). Seawater samples were collected at selected stations to analyze the full suite of ventilation tracers, including CFCs, SF$_6$, ^{39}Ar, ^{14}C, as well as the anthropogenic radionuclides ^{129}I and ^{236}U. The transect covered the Nansen and Amundsen Basins, where both radionuclides had already been measured in 2015 [1], and the area north of Greenland where no data on ^{129}I and ^{236}U was available before (Fig. 1).

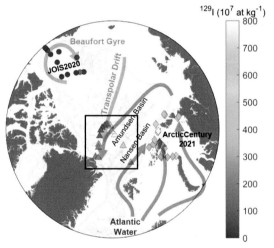

Fig. 1: *^{129}I concentrations in the surface layer of the Arctic Ocean in 2020/21. Stations from the Arctic Ventilation expedition are outlined by the black box. Arrows represent surface waters of Pacific (blue) and Atlantic (red) origin.*

As highlighted by the ^{129}I concentrations (Fig. 1), the cruise track crossed the interface between Atlantic- and Pacific-origin waters in the surface layer. Higher ^{129}I concentrations in the Nansen and Amundsen basins reflect the tracer signal carried by Atlantic Water, originating from discharges from the European nuclear reprocessing plants. This is also evident from the samples collected upstream in the Barents Sea as part of another expedition [2]. Pacific Waters are

characterized by a low ^{129}I tracer signal, also seen in samples from the Canada Basin taken in 2020 (JOIS2020).

A comparison of ^{129}I and ^{236}U data from one station in the Amundsen basin sampled in 2021 to data from a station close-by visited in 2015 showed slight differences between the years, especially in the upper water column and around 1000 m (Fig. 2). While ^{129}I increased in the surface layer, ^{236}U showed a decrease in the upper 500 m. However, sampled depths and depth resolution were not exactly the same in both years, which hampers a straight comparison.

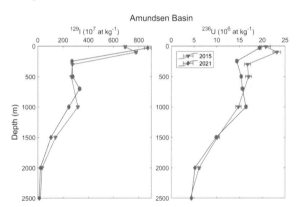

Fig. 2: *Profiles of ^{129}I (left) and ^{236}U (right) from the Amundsen basin comparing 2015 (blue) to 2021 (red).*

As a next step, the new data on ^{129}I and ^{236}U will be used to calculate transit time distributions, allowing to constrain circulation times [3] and to estimate anthropogenic Carbon inventories.

[1] N. Casacuberta et al., J. Geophys. Res. Oceans 123 (2018) 6909

[2] A.-M. Wefing et al., LIP annual report (2022) 68

[3] A.-M. Wefing et al., Ocean Sci. 17 (2021) 111

[1]Environmental Systems Sciences, ETH Zurich

PRELIMINARY RESULTS FROM JOIS 2020 & 2021

First insights into water flow paths and timescales using ^{129}I and ^{236}U

A. Payne[1], A.-M. Wefing[1], N. Casacuberta[1], M. Christl, C. Vockenhuber

In 2020 120 samples were collected for ^{236}U and ^{129}I as part of the JOIS (Joint Ocean Ice System Survey) expedition to the Canada Basin, the first time the two isotopes had been extensively sampled in tandem in the region (Fig 1.). Both isotopes are released from nuclear-reprocessing plants in Europe, labeling Atlantic water (AW) and allowing the study of flow pathways, and timescales [1] and quantifying heat transport to the basin.

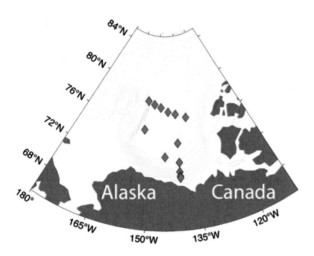

Fig 1. *Location of stations sampled during JOIS 2020 expedition.*

The station profiles (Fig. 2) show that the two radionuclides clearly label the AW that sits between 200 m and 2000 m depth. Overlying Pacific sourced water carries the low-level global fallout signal, while the deeper Canada Basin Deep and Bottom water has no tracer signal due to its age and relative isolation from the upper Atlantic layer. Initial Transit Time Distribution (TTD) calculations [2] indicate that the average time taken for water to reach the basin is around 30 yr, however, there are latitudinal differences seen across the basin and between the two branches within the AW that need to be investigated further.

In 2021 additional 120 ^{129}I samples were collected. For stations that were repeated that year, a significant increase in the signal carried in the AW was seen. Future work (with samples collected during the 2022 JOIS expedition) will allow us to begin disentangling the drivers behind these temporal changes, which may include the response of surface and mid-depth currents to atmospheric forcing.

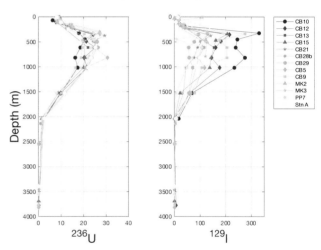

Fig 2. *Depth profiles of ^{236}U and ^{129}I concentrations for all stations sampled during JOIS 2020. Note concentrations for ^{236}U are x 10^{-6} at L^{-1}, while 129I concentrations are x 10^{-7} at L^{-1}.*

[1] N. Casacuberta et al., JGR: Oceans. 123 (2018) 6909

[2] A.-M. Wefing et al., Ocean Sci. 17 (2021) 111

[1]Environmental Systems Science, ETH Zurich

2022 JOIS EXPEDITION

Identifying water mass distributions in the western Arctic Ocean

A. Payne[1], A.-M. Wefing[1], B. Williams[2], M. Christl, N. Casacuberta[1]

The Canada basin is the meeting point of Pacific and Atlantic-derived waters in the Arctic, each contributing water to the basin with vastly different physical and tracer properties. Due to the basin's extreme remoteness, and with access limited to the summer months, many questions remain about the distribution of waters within the basin, and their temporal evolution in response to climate forcing. The JOIS / BGOS (Joint Ocean Ice Survey / Beaufort Gyre Observing System Survey) expedition is undertaken annually in the Canada basin, as a joint effort between the Department of Fisheries and Oceans (Canada) and Woods Hole Oceanographic Institute (USA).

Fig 1: *The CCGS Louis St. Laurent in sea ice during deployment of an Ice Tethered Profiler at 78 °N.*

2020 was the first year in which a dual-tracer approach using ^{236}U and ^{129}I to identify the different sourced waters in the basin was employed. Atlantic-derived water carries a nuclear reprocessing signal released from plants in Europe, while Pacific waters carry the low-level global fallout signal from nuclear weapon testing. The 2022 cruise builds on the initial study by expanding the sampling area and increasing the resolution from 120 to 280 samples for both isotopes. The dual-tracer method allows us to not only identify water masses, but by using the Transit Time Distribution method (TTD), we can estimate the ages of these waters. New data complements the time-series in this region that builds on previously collected ^{129}I, allowing us to identify changes in the flow paths and/or ages.

80 additional samples were collected for ^{14}C analysis which will be carried out at ETH, and 27 samples for ^{39}Ar which will be analyzed at Uni Heidelberg. The combination of these two tracers will allow the estimation of TTDs in the deepest waters of the Canada Basin, which are too old to be labeled by the anthropogenic radionuclides. A further 20 samples were also collected from the surface waters to explore the potential of lithogenic Hafnium and Neodymium as tracers to identify the numerous different components present in the compositionally complex uppermost waters of the basin.

Fig. 2: *Deployment of the CTD Rosette used for sampling.*

---

[1]*Environmental Systems Science, ETH Zurich*
[2]*Ocean Sciences, Department of Fisheries and Oceans, Sidney, Canada*

TRACING OCEAN CIRCULATION AROUND ICELAND

Using ^{236}U and ^{129}I to identify water mass provenance and mixing

D. Dale[1], M. Christl, A. Macrander[2], S. Ólafsdóttir[2], R. Middag[3], N. Casacuberta[1]

Iceland stands at an important gateway where Arctic and Atlantic waters interact. Atlantic waters pass northward and circulate in the Arctic before returning southward in the East Greenland Current (EGC). Some is cooled and sinks to contribute to overflows of the Iceland-Scotland Ridge such as Denmark Strait Overflow Waters (DSOW). These are key processes in Arctic warming and deep ocean ventilation.

This system has been tagged with anthropogenic radionuclides ^{129}I and ^{236}U by bomb tests in the 1950-60s and point-source nuclear reprocessing plants (NRPs) at Sellafield (UK) and La Hague (FR) since the 1960s, providing an opportunity to trace the origins of water masses in the region and their transit timescales [1].

We collected 200 seawater samples from around Iceland during two cruises in 2021: "Iceland21" and the GEOTRACES cruise "MetalGate" (Fig. 1). Plotting ^{236}U and ^{129}I concentrations for water masses enables inference of mixing between masses and estimates of mixing volumes (Fig. 2).

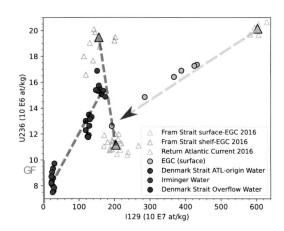

Fig. 2: ^{236}U-^{129}I plot of samples from this (circles) and previous (triangles) studies [2]. Filled triangles = mean. colours per masses in Fig. 1.

The surface-EGC (green) is diluted with Irminger Water (IW) (purple) or water with a global fallout (GF) signal, but mostly not until close to Iceland. The shelfbreak-EGC (pink) mixes approx. 60-40 with Return Atlantic Current (yellow) to form the Atlantic-origin layer in Denmark Strait (red). DSOW (blue) forms from a mix of this layer and IW after cooling and sinking north of Iceland. Our results provide geochemical evidence for Atlantic water "shortcutting" through the Nordic Seas.

[1]   N. Casacuberta et al., J. Geophys. Res. Oceans 123 (2018) 6909

[2]   A.-M. Wefing et al., Ocean Sci. 17 (2021) 111

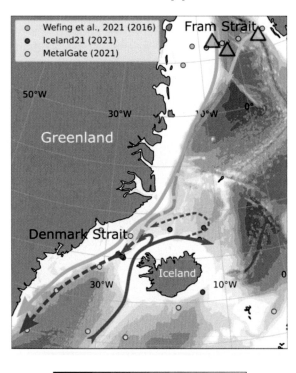

Fig. 1: *Schematic of ocean circulation in the Nordic Seas with location of samples used in this study. See Fig. 2 for explanation of triangles.*

[1]*Environmental Systems Sciences, ETH Zurich*
[2]*MFRI, Hafnarfjörður, Iceland*
[3]*NIOZ, Ocean Systems, Texel, Netherlands*

^{129}I IN THE GULF OF SAINT LAWRENCE (CANADA)

A first insight on the radionuclide distribution

L. Raimondi[1], C. Vockenhuber, N. Casacuberta[1]

Whereas most of the anthropogenic radionuclide ^{129}I is found in the Norwegian coast and Arctic Ocean as a result of its discharge by European nuclear fuel reprocessing plants (RPs; i.e. Sellafield and La Hague), detection of this isotope extends far beyond these regions. Indeed, ^{129}I is transported into the subpolar North Atlantic (SPNA) through the Fram and Davis Straits and further south to the subtropical North Atlantic via the Deep Western Boundary Current [1, 2].

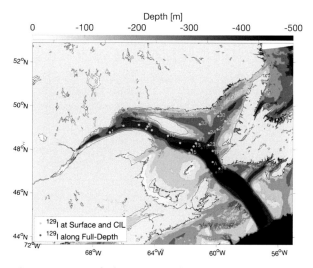

Fig. 1: *Map of the GSL with locations of TReX stations (in red).*

The Gulf of Saint Lawrence (GSL, Fig. 1) is a semi-enclosed sea tightly connected to the SPNA through the Strait of Belle-Isle in the northeast (that connects it to the Labrador Sea) and the Cabot Strait in the south. It is characterized by large riverine input from the Saint Lawrence River (SLR) and by intense naval traffic and is therefore heavily anthropogenically impacted.

The connection of the GSL to the SPNA tagged with RPs' ^{129}I as well as its large riverine discharge with potential radionuclide input from Canadian reprocessing plants along the SLR watershed, make this region particularly interesting for an exploratory study of ^{129}I.

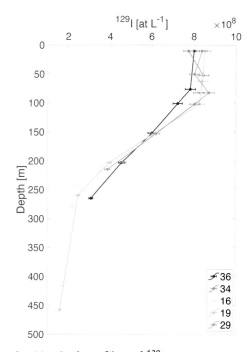

Fig. 2: *Vertical profiles of ^{129}I concentrations in the GSL.*

During the TReX Program, specifically designed to understand the dispersal of contaminants within this region, 70 samples for the measurements of ^{129}I were collected during a purposeful tracer release experiment.

Preliminary results (Fig. 2) show that the region has already received waters tagged with the RPs' ^{129}I signal particularly in the surface and intermediate layers. On the other hand, despite the relatively shallow bathymetry (~500m), the deeper GSL shows values of ^{129}I that suggest only a global fallout source. Further analysis is required to better understand the distribution of this tracer in the region.

[1] N. Casacuberta and J. N. Smith, Annu. Rev. Mar. Sci. 15 (2023) 203
[2] J. N. Smith et al., J. Geophys. Res. Oceans 121 (2016) 8125

[1]Environmental Systems Sciences, ETH Zurich

2022 EXPEDITIONS TO LABRADOR SEA AND DAVIS STRAIT

Sampling ^{129}I and ^{236}U samples from Arctic and Atlantic origin

L. Leist[1], C. Lee[2], K. Azetku[3], M. Ringuette[3], K. Kündig[1], M. Christl, N. Casacuberta[1]

The Labrador Sea is an important region for intermediate and deep-water formation and connects the Arctic Ocean with the Subpolar North Atlantic. With the analysis of the tracer pair ^{129}I and ^{236}U at the northern (Davis Strait) and southern border of the Labrador Sea (AR7W) we hope to learn more about water mass formation, origin, and transport timescales.

Previous studies already showed a general increase of the ^{129}I concentrations along the historic AR7W line sampled since 1993 [1, 2, 3]. In the past the sampling mainly focused on the deep and intermediate waters and only in 2020 ^{236}U was sampled for the first time. From 2020, sampling includes the tracer pair ^{129}I and ^{236}U, and further extends the previous sampling plan to both inflow and outflow of different surface and coastal currents. Results will provide new insights to the formation of Labrador Sea Water.

During the AR7W Cruise 80 ^{129}I and ^{236}U samples as well as 116 ^{14}C samples were collected by K. Kündig (Fig. 1) and sent to ETH Zurich.

Fig. 1: *K. Kündig at RV Atlantis at departure for the AR7W cruise in the Labrador Sea in May 2022.*

The Davis Strait is the north boundary of the Labrador Sea and a gateway from the Baffin Bay and the Arctic to the Labrador Sea. During the Davis Strait Cruise in September-October 2022 three lines were sampled: one crossing the Davis Strait, one reaching into the Baffin Bay and one in the very north of the Labrador Sea (Fig. 2). This was the first time that Davis Strait was sampled for the artificial isotopes ^{129}I and ^{236}U.

The focus of the cruise was to take samples from (i) the West Greenland Current and its evolution into the Baffin Bay; (ii) Arctic waters coming from the Baffin Bay exiting the Arctic along the shelf of Baffin Island. The later ones may contribute to the formation of the Labrador Current in the Labrador Sea. Again, the sampling of the tracer pair will add to the understanding of water mass origin, mixing and time scales of transport.

Fig. 2: *L. Leist sampling close to Baffin Island during the Davis Strait Cruise in Oct. 2022.*

A total of 164 ^{129}I and ^{236}U samples and 40 ^{14}C samples were sampled during that cruise and sent to ETH for further analysis.

[1]   M. Castrillejo et al., Biogeosciences 15 (2018) 5545

[2]   M. Castrillejo et al., Front. Mar. Sci. 9 (2022) 897729

[3]   J. N. Smith et al., J. Geophys. Res. Oceans 110 (2005) C05006

---

[1]*Environmental Systems Science, ETH Zurich*
[2]*Woods Hole Oceanographic Institution, USA*
[3]*Bedford Oceanography, Dartmouth, Canada*

^{129}I AND ^{236}U FROM BOCATS AND AR7W EXPEDITIONS

Studying water masses and their mixing in the Labrador Sea and SPNA

L. Leist[1], N. Casacuberta[1], M. Christl, C. Vockenhuber, J. Smith[2], M. Castrillejo[3]

The Labrador Sea and the Subpolar North Atlantic (SPNA) are important regions of intermediate and deep-water formation, which carry current climate properties of surface waters to the depth of the Ocean. The SPNA is composed by warm waters that flow northwards to higher latitudes and by deeper and colder waters coming from the north and circulating southwards [1,2]. Here we present an update on the results of ^{129}I and ^{236}U collected during the AR7W (Labrador Sea, from Canada to Greenland) and BOCATS (SPNA, from Portugal to Greenland) expeditions in 2020 and 2021, respectively (Fig. 1).

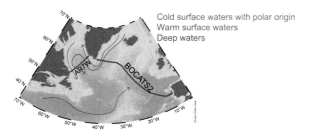

Fig. 1: *Map of the Labrador Sea and SPNA. AR7W line between Canada and Greenland and BOCATS line between Portugal and Greenland.*

The combination of both cruises (Fig. 2) provides a comprehensive overview of the ^{129}I concentrations in seawater at key regions of intermediate and deep-water mass formation.

Results of ^{129}I concentrations show a general increase from east (Portugal) to west (Labrador Sea) and to higher latitudes. This was expected since the sources of these isotopes, the Nuclear Reprocessing plants in Sellafield and La Hague, are upstream of the sampling locations. Closer to Greenland and Canada, waters are mainly influenced by southward flowing arctic waters. In the Arctic, previous studies showed that waters from Atlantic origin are tracer labelled and therefore carry a higher signal of both ^{129}I and ^{236}U. Especially high tracer signals (up to 250 x10^7 at/kg) were found in advective currents

along the Greenlandic and Canadian shelf, and at the bottom water of the Labrador Sea (Fig. 2).

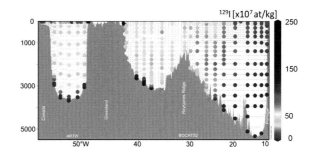

Fig. 2: ^{129}I *concentrations at the AR7W and BOCATS transect.*

Concentrations of 30 x10^7 at/kg intermediate depths in the in the Labrador Sea and the Icelandic Basin (Greenland to Reykjanes Ridge) are surprisingly high and may trace the recirculation of waters in these regions for multiple years, up to decades. They further could give insights in the formation process via winter convection [3].

On the eastern side of the Reykjanes Ridge only the Iceland-Scotland overflow water is showing elevated concentrations, while the deep and bottom waters with origin in the southern ocean do not carry any tracer signal, proving the older age of these waters.

Results on ^{236}U are not shown here, because the sampling resolution was not as high as for ^{129}I. Still, its distribution pattern is similar.

[1] M. Castrillejo et al., Biogeosciences 15 (2018) 5545

[2] M. Castrillejo et al., Front. Mar. Sci. 9 (2022) 897729

[3] K. L. Lavender et al., Deep-Sea Research I 52 (2005) 767

^{1}*Environmental Systems Science, ETH Zurich*
^{2}*Bedford Oceanography, Dartmouth, Canada*
^{3}*Imperial College London, UK*

^{129}I AND ^{236}U IN THE SUBTROPICAL NORTH ATLANTIC

Results from the BATS expedition in December 2021

A.-M. Wefing[1], N. Casacuberta[1], C. Vockenhuber, M. Christl, K. Kündig[1]

In December 2021, seawater samples for ^{129}I and ^{236}U analysis were taken in the subtropical North Atlantic, at the BATS (Bermuda Atlantic Time-Series Study) station as well as stations 4 and 6 of the validation line (BVAL) further south [1]. The aim of this sampling program was to create a time series of ^{129}I and ^{236}U close to Bermuda, following up on the first sample collection in 2019 [2]. The stations are located within the Deep Western Boundary Current and its interior branches, and potentially also influenced by Antarctic Bottom Water (AABW) (Fig. 1).

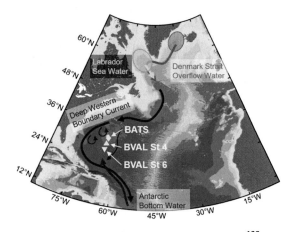

Fig. 1: *Locations of stations sampled for ^{129}I and ^{236}U in 2021 and major deep currents.*

Full-depth profiles of ^{129}I and ^{236}U concentrations measured at the three stations showed elevated levels of both radionuclides at around 1500 m and around 3500-4000 m depth (Fig. 2). Concentrations are highest at the BATS station and decrease southwards along the validation line (BVAL St 4 and 6). The peaks at the two depths are explained by the presence of Labrador Sea Water (LSW, around 1500 m) and Denmark Strait Overflow Water (DSOW, 3500-4000 m) (Fig. 1), carrying the radionuclide signal from northern latitudes.

For ^{129}I, concentrations in DSOW at the BATS station increased significantly from 2019 to 2021 (Fig. 2), suggesting that the strong rise in the discharge history from Sellafield and La Hague had reached the subtropical North Atlantic.

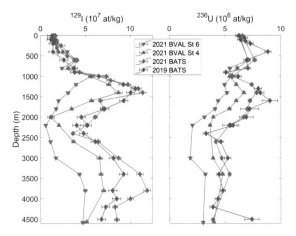

Fig. 2: *Profiles of ^{129}I (left) and ^{236}U (right) concentrations at the three stations sampled in 2021 (locations in Fig. 1), compared to concentrations from 2019.*

For ^{236}U, elevated concentrations were also found in the upper 100 m, pointing to the atmospheric input from nuclear weapon tests in the 1950s and 60s. In the deepest samples concentrations of both radionuclides were significantly above natural background levels that would be expected for the extremely old AABW. This suggests that no pure AABW was present at these latitudes in 2021.

The time series of ^{129}I and ^{236}U at the BATS station will be used to revisit the pathways, timescales, and missing properties of the Deep Western Boundary Current and its interior branches.

[1] A.-M. Wefing et al., LIP annual report (2021) 82

[2] N. Casacuberta et al., LIP annual report (2020) 83

[1]*Environmental Systems Sciences, ETH Zurich*

FIRST ^{129}I MEASUREMENT ON MILEA

Determination of ^{129}I/^{127}I ratio in environmental samples

M. Mindová[1], M. Daňo[1], M. Němec[1], C. Vockenhuber, X. Hou[2]

AMS measurements of ^{129}I on MILEA are still not fully routinely established. Measurement of ^{129}I in artificial and real environmental samples, and intercomparison among different AMS laboratories is crucial for optimization and standardization of the instrument and measurement method, as well as for its analytical accuracy. Initial tests show excellent performance of MILEA for ^{129}I AMS [1]. Hence, the measurement of ^{129}I in AgI target prepared from environmental and testing samples was planned on two MILEA systems at LIP, ETH Zürich and at RAMSES laboratory, Rez, Czech Republic. The first ^{129}I measurements of environmental samples on the MILEA at ETH are presented here.

Firstly, low-background commercially available materials, namely 'Solca' iodine-rich borehole water from Czech Republic, potassium iodide (KI), caesium iodide (CsI), were measured together with iodine reagent from Woodward company (WWI). The measured values (Tab. 1) are up to one order of magnitude higher than the lowest value reported values for Woodward iodine [1, 2]). Despite this fact, these materials can be possibly used for beamline tuning and can serve as blank materials for high ratio samples, with advantage of their commercial availability and easy preparation.

	^{129}I/^{127}I $[10^{-14}]$	Unc. [%]
WWI	7.3	5.8
Solca	15.6	2.4
KI	17.7	2.1
CsI	15.2	2.5

Tab. 1: *^{129}I/^{127}I atomic ratios measured in commercially available low-background materials together with Woodward iodine blank.*

A batch of seawater samples collected in Danish waters in recent years were also analysed. Inorganic iodine in these samples was first separated using solvent extraction and precipitated as AgI at DTU, then mixed with Nb powder and pressed in the Cu target holders, and measured with AMS MILEA at ETH. The resulting ^{129}I/^{127}I ratios range from 1×10^{-7} to 2×10^{-6} (Fig. 1). A much higher ^{129}I level (10^{-6}) was observed in the North Sea waters compared to those collected in the Baltic Sea (10^{-7}) indicating the dominant source of ^{129}I in seawater in this region is the marine discharges from the reprocessing plants at Sellafield and La Hague. The contaminated water from southern North Sea moves along the European coast to the North and a small branch of it enters the Baltic Sea. The results will be used in oceanographic tracer studies. It is planned to remeasure some of those samples at the MILEA at RAMSES laboratory for performance comparison.

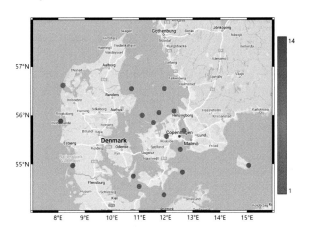

Fig. 1: *^{129}I/^{127}I $[10^{-7}]$ atomic ratios for respective seawater sampling points around Denmark.*

[1] C. Vockenhuber et al., LIP annual report (2022) 19

[2] L. K. Fifield et al., NIM-B, 530 (2022) 8

[1]*Nuclear Chemistry, CTU Prague, Czech Republic*
[2]*Environmental and Resource Engineering, DTU, Lyngby, Denmark*

ANTHROPOGENIC U IN THE SOUTHERN ATLANTIC OCEAN

^{233}U and ^{236}U in a sediment core from the anoxic shelf of Namibia

E. Chamizo[1], M. Christl, P. Gautschi, M. López-Lora[2]

The Benguela Upwelling System, along the Namibian coast in the Southern Atlantic Ocean, is one of the four major coastal upwelling ecosystems of the world. Organic rich, sub or anoxic sediments from this region contain high quantities of natural U. Under oxygen depleted conditions, soluble U(VI) is reduced to poorly soluble U(IV) which is then incorporated into the sediments (authigenic U) [1]. In this work, a sediment core collected in 2015 at 116 m depth at the anoxic shelf of the Namibian coast (25°00.000'S, 14°28.200'E, labeled as 25020) was analyzed for its ^{236}U/^{238}U and ^{233}U/^{238}U atomic ratios (ARs) at the 300 kV MILEA ETH AMS facility.

21 samples corresponding to the 0-31 cm depth range (2.5 g aliquots) were chemically processed at CNA. They were dissolved using concentrated HNO_3, HF and H_2O_2. Actinides were concentrated with $Fe(OH)_3$, and U (and Pu) subsequently purified using UTEVA® (and TEVA®) resins. Any spike or Fe carrier was added to prevent contamination. Abundance sensitivities for "^{239}Pu"/^{238}U ("^{236}U"/^{235}U) and "^{233}U"/^{232}Th ARs below the 10^{-12} level were demonstrated during the measurements.

The obtained ^{236}U/^{238}U and ^{233}U/^{238}U ARs range from 1×10^{-12} to 300×10^{-12} and from 0.05×10^{-12} to 10×10^{-12}, respectively (Fig. 1). The deepest samples (from 25 to 31 cm depth) produced ^{238}U^{3+} beam currents in the 15-45 nA range, due to their unusually high ^{238}U content (in the 50-100 μg/g interval as reported in [2]). ^{236}U/^{238}U ARs at the 10^{-12} level with 5-10% statistical uncertainties were reached for those highly U enriched samples. ^{233}U count rates in that depth range were close to the instrumental blank levels, which explains the large error bars (Fig. 1).

Our results show a gradual increase in the ^{236}U/^{238}U ARs starting at 25 cm depth, reaching the maximum value at surface. In contrast, a structure in the ^{233}U/^{238}U AR depth profile is clearly observed above 10 cm, with distinguishable peaks at 5.5 and 1.5 cm depth and ^{233}U/^{236}U ARs in the $(0.4\text{-}3) \times 10^{-2}$ range (i.e. the estimated average value for global fallout for the Northern Hemisphere is $(1.40\pm0.15) \times 10^{-2}$ [3]). This might indicate different timing signals of both ^{233}U and ^{236}U at the studied site. Note that no other source than the atmospheric testing of nuclear weapons is expected in the Southern Atlantic Ocean.

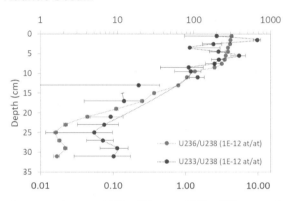

Fig. 1: *Obtained ^{233}U/^{238}U and ^{236}U/^{238}U ARs (in 10^{-12} at/at units) for the 25020 sediment core collected at the anoxic shelf of Namibia.*

These are the first ^{233}U and ^{236}U results for a sedimentary reservoir from the Southern Atlantic Ocean. Ongoing work on the Pu isotopic composition would give more information on the actinides sources to the region. The authors are grateful to IAEA Environment Laboratories in Monaco for providing the measures samples.

[1] R. F. Anderson et al., Geochim. Cosmochim. Acta 53(9) (1989) 2215
[2] M. L. Abshire et al., Earth Planet. Sci. Lett. 529 (2020) 115878
[3] K. Hain et al., Nat. Comm. 11 (2020) 1275

[1]Centro Nacional de Aceleradores, Seville, Spain
[2]Health, Medicine and Caring Sciences, Linköping Univ., Sweden

REACTOR-DERIVED VS ENVIRONMENTAL PLUTONIUM

Separating trace environmental plutonium in soils using ^{244}Pu/^{239}Pu

A. Madina[1], M. Christl, P. Gautschi, F. Livens[2], K. Semple[1], C. Tighe[1], M.J. Joyce[1]

Reactor-derived plutonium can augment environmental arisings from being 100% weapons fallout, e.g., Salzburg (Chernobyl) [1], Enewetak (incomplete fission) [2] and the UK Lake District (Windscale) (Fig 1).

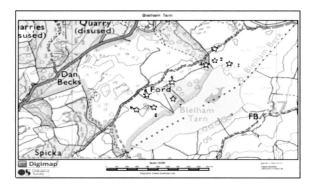

Fig. 1: *Sample locations in the Lake District.*

Whilst this contribution can be gauged in extraordinary cases where ^{240}Pu/^{239}Pu is distinct from fallout (as for fast breeders [3]), this is rarely the case for thermal reactors where the ranges in ^{240}Pu/^{239}Pu can overlap. However, ^{244}Pu *is* a distinction of fallout we can use, adapting the ^{240}Pu relationship [3], as per,

$$F_{L_{239}} = \frac{\left(R_{244/239} - R_{G_{244/239}}\right)}{\left(R_{L_{244/239}} - R_{G_{244/239}}\right)}$$

where $F_{L_{239}}$ is the local ^{239}Pu proportion, $R_{244/239}$ is the ^{244}Pu/^{239}Pu measurement, $R_{L_{244/239}}$ is the ratio for local material and $R_{G_{244/239}}$ is the global fallout average.

As, for Irish Sea sediments [1], $R_{L_{244/239}} \sim 3 \times 10^{-6}$ and using Lake Erie (Canada) data for $R_{G_{244/239}} = (14.4 \pm 2.8) \times 10^{-5}$ [4], $R_{L_{244/239}} \ll R_{G_{244/239}}$ so,

$$F_{L_{239}} \sim 1 - \frac{R_{244/239}}{R_{G_{244/239}}}$$

The % reactor-derived ^{239}Pu on this basis for the Lake District, Salzburg [1] and Enewetak [2] are given in Table 1 (for Enewetak $R_{G_{244/239}} = (11.8 \pm 0.7) \times 10^{-4}$ for Ivy Mike [5] is used, being southern hemisphere). This, with local ^{240}Pu/^{239}Pu = (0.03 ± 0.03), suggests the 98% ^{239}Pu at Enewetak is unfissioned ^{239}Pu from the Quince safety test [2]. Similarly, for the Lake District, local ^{240}Pu/^{239}Pu $(0.21 \pm 0.06) \sim$ Irish Sea sediments [1] (0.2100 ± 0.0004) but inconsistent with material from the fire-affected zone of the destroyed reactor (^{240}Pu/^{239}Pu $\sim$0.0217) [6]. For Salzburg, local ^{240}Pu/^{239}Pu of (0.18 ± 0.02) is consistent with northern Europe, i.e., (0.1814 ± 0.0006) [7] but not Chernobyl ^{240}Pu/^{239}Pu (0.41 ± 0.02) [8]. Hence, Lakeland comprises locally derived ^{239}Pu but not from Windscale, and Salzburg appears devoid of a local contribution, despite prior reports [1].

Location	$R_{244/239}$ / $\times 10^{-5}$	$F_{L_{239}}$ / %
Lake District	11.9 ± 2.8	18 ± 4
Salzburg [3]	5.7	60 ± 12
Enewetak [2]	3.2×10^{-5}	98 ± 21

Tab. 1: $F_{L_{239}}$ for this work, compared with [2].

[1]   P. Steier et al., NIM-B 294 (2013) 160
[2]   T. F. Hamilton et al., J. Env. Radio. 242 (2022) 106795
[3]   C. Tighe et al., Nat. Comm. 12 (2021) 1381
[4]   C. Wendel et al., Env. Int. 59 (2013) 92
[5]   H. Diamond. Phys. Rev. 119 (1960) 2000
[6]   D. G. Pomfret, Japan Health Physics Society, Tokyo., Int. Cong. Int. Radiation Protection Association (2000)
[7]   J. M. Kelley et al., Sci. Total Environ. 237/238 (1999) 483
[8]   T. Warneke et al., EPSL 203 (2002) 1047

[1]*Engineering, Lancaster Univ., UK*
[2]*Chemistry, Univ. Manchester, UK*

COMBINING AMS AND DGT FOR ACTINIDE MEASURES

An overview of results from the latest LIP and CHUV-IRA collaboration

J. Chaplin[1], M. Christl, P. Froidevaux[1], M. Straub[1]

This year saw the conclusion of another collaborative project between LIP and the Institute of Radiation Physics at Lausanne University Hospital (CHUV-IRA). Within the framework of a Ph.D. thesis, new DGT (diffusive gradients in thin-films) samplers (Fig. 1) were developed to be able to capture actinides (U, Pu, Am and Cm) from complex solutions [1]. AMS was implemented to measure actinides from the samplers and other environmental samples. We deployed these samplers over two two-week campaigns in the UK's Esk Estuary, which is contaminated with legacy discharges from the nearby Sellafield nuclear facility.

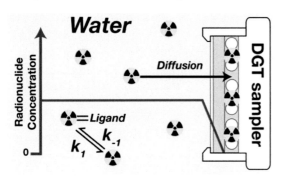

Fig. 1: *The passive diffusion principles of the DGT technique for measuring actinides in water.*

Overall, 410 isotope-to-tracer ratio data points were produced at LIP resulting from the two sampling campaigns. These mainly involved Tandy to measure ^{233,236}U, $^{239-242}$Pu and ^{241,243}Am, with some ^{236}U measures performed using MILEA. By combining the data produced from the DGT samplers in the estuary water with the concentration data in the biota and ultrafiltered estuary water, we were able to assess DGT as a bioavailability proxy [2]. As a time-integrated *in-situ* sampling technique, DGT is possibly more valid than spot-sampled water at predicting the uptake of these actinides in biota [2].

We also deployed DGT samplers in the adjacent estuary sediments. In sediments, the DGT samplers act like a plant root, since they function on the basis of passive diffusion to an elemental sink. We were therefore able to assess the remobilisation of bioavailable actinides from the sediment [3]. We observed here that the isotopic ratios of sediment-associated plutonium matched that of the biota rather than that of the seawater, indicating that the sediments are now the principal source of supply to the biota [3]. Critically, this indicates that the cessation of nuclear discharges to the environment do not necessarily stop radionuclide supply to biota.

We were also able to calibrate the DGT samplers for deployments in a spent nuclear fuel pool, using TANDY to additionally test for Cm [4]. We delivered fuel pool concentration data to a nuclear facility and produced isotopic reports. These allowed us to accurately date a leaking fuel rod [4], providing the first known implementation, and preliminary validation, of a recently developed method for dating fuel rods based on their Cm-isotope signature [5].

We're happy to report that following the conclusion of the work of this Ph.D. thesis, the Swiss Federal Nuclear Safety Inspectorate is implementing these samplers for future work, including the monitoring of the Aare waters.

[1] J. Chaplin et al., Anal. Chem. 93 (2021) 11937

[2] J. Chaplin et al., ACS ES&T Water 2 (2022) 1688

[3] J. Chaplin et al., Water Res 221 (2022) 118838

[4] J. Chaplin et al., ACS Omega 7 (2022) 20053

[5] M. Christl et al., J. Radioanal. Nucl. Chem. 322 (2019) 1611

[1]*Radiophysique, Centre Hospitalier Universitaire Vaudois, Lausanne*

FLUORIDE TARGET MATERIALS FOR ^{239}Pu ANALYSES

Testing of fluoride matrices for ^{239}Pu measurement with MILEA system

K. Fenclová[1], T. Prášek[1], M. Němec[1], M. Christl, P. Gautschi

The aim of this research was to evaluate the performance of prepared plutonium fluoride (^{239}Pu + PrF$_3$) and oxide matrices (^{239}Pu + Fe$_2$O$_3$) with the MILEA AMS system at LIP, ETH Zürich. The measurements included tuning the system for ^{239}Pu detection using fluoride and oxide target materials, background analyses carried out by scanning both HE ESA and second HE magnet, focusing on ^{238}U interference, and also ionization efficiency tests for plutonium negative ions ^{239}PuF$_4^-$, ^{239}PuF$_5^-$ and ^{239}PuO$^-$.

The contribution of ^{238}U to background was investigated by measuring both the plutonium fluoride and oxide material along with samples doped with different amounts of uranium, which allows for direct comparison of the influence of different matrices on the background. Fig. 1 represents an example of ^{239}Pu background measurement, showing the HE2 magnet scans of plutonium fluoride (^{239}PuF$_4^-$).

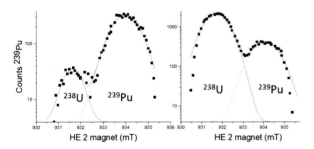

Fig. 1: *MILEA background measurements – HE2 magnet scans of plutonium fluoride (left) and plutonium fluoride with added 100 µg of ^{238}U.*

The left HE2 magnet spectrum in Fig. 1 coresponds to plutonium fluoride material (^{239}PuF$_4^-$) without any addition of uranium. The number of ^{239}Pu counts is 12 times higher than for ^{238}U, interfering plutonium peak with 12 % overlap. The right spectrum shows the HE2 magnet scan where 100 µg of ^{238}U was added to the sample. This lead to a substantial increase of ^{238}U counts, resulting in the peak overlap as high as 46 %.

The measurements of the overall detection efficiency were performed for the samples of plutonium fluoride matrix admixtures with PbF$_2$ (1:1 mass ratio) and plutonium oxide matrix admixtures with Nb powder (1:1). The efficiency values were calculated as a ratio of ^{239}Pu^{3+} atoms counted by GID and the total number of plutonium atoms in the cathode which was the same for both materials. The fluorides were sputtered about 6 hours and oxides 12 hours until negligible GID signal values. The comparison of these three injected masses is displayed in Fig. 2, relatively to PuO$^-$ extracted from plutonium oxide. The highest yield values were obtained for the plutonium fluoride sample material while the PuF$_4^-$ ion was injected to accelerator and Pu^{3+} ions were detected by GID. This efficiency was twice as high as the value obtained by injecting PuO$^-$ while almost negligible efficiency was achieved when PuF$_5^-$ ions were injected.

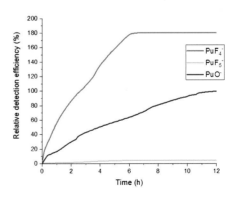

Fig. 2: *Comparison of detection efficiency for fluorides (PuF$_4^-$/PuF$_5^-$) and oxides (PuO$^-$). The detection efficiencies values are reported relatively to the PuO$^-$ detection efficiency.*

---

[1]*Nuclear Chemistry, CTU Prague, Czech Republic*

SINGLE-SAMPLE ANALYSIS OF ACTINIDE ISOTOPES

Setting up a chromatographic separation system for U, Pu and Am

T. Prášek[1], M. Christl, H. Pérez Tribouillier

Multi-isotope analysis of actinides within a single environmental sample, usually seawater, is generally a challenging process, mainly due to their isobaric overlaps and tailing effect of neighboring masses occurring during AMS measurement. To obtain reliable results, highly efficient mutual chemical separation is required prior to instrumental analysis, which is particularly important for ^{239}Pu, affected by significantly more abundant ^{238}U tailing, and ^{241}Am sharing the mass with ^{241}Pu.

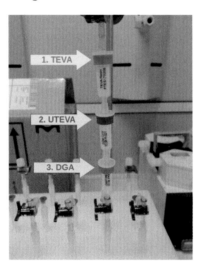

Fig. 1: *Sequence of chromatographic columns for selective sorption of Pu, U and Am. Optimal flow rate of about 1 ml/min is achieved by using an in house made vacuum box system.*

For this purpose, a chromatographic system has been set up, based on separation procedure designed at IRA Lausanne. The system utilizes a consecutive sorption of plutonium, uranium and americium, pre-concentrated with iron hydroxide co-precipitation, on three stacked columns, containing TEVA®, UTEVA® and DGA® resins, respectively. Adsorption occurs in the environment of 8M nitric acid, with Pu being adsorbed selectively on the first, U on the second and Am (along with Cm) on the third column.

Subsequent elution is carried out separately with different solutions. Particularly for Pu, different oxidation states have to be considered during the separation. Its sorption occurs in IV+ state, achieved by addition of sodium nitrite to the initial solution, whereas the elution step makes use of reduction to III+ with hydroxylamine, which needs to be destroyed with concentrated nitric acid before further processing of the sample.

Fig. 2: *Oxidative destruction of hydroxylamine in Pu fraction – a violent reaction occurs after sufficient amount of nitric acid is added.*

Initial tests of the separation efficiency achieved with the setup have been performed with model samples containing isotopic spikes ^{233}U, ^{242}Pu and ^{243}Am in 2 litres of acidified Milli-Q water. After the separation steps, samples for AMS analysis have been prepared for each fraction following standard iron-oxide procedure and analysed with MILEA system. Measurement results suggest promising separation efficiency of all 3 elements and will be further studied with standardized seawater samples.

---

[1]*Nuclear Chemistry, CTU Prague, Czech Republic*

MATERIALS SCIENCES

View into the open RBS chamber with detectors for RBS, He-ERD and PIXE

A new sample changer for IBA analyses

Novel gases for gas ionization chambers

Elemental depth profiles of weathered polymers

Reaction front control in Ni/Al RM

Lithicone coatings for Li-ion batteries

Composition of anodic aluminum oxides

Origin of irradiation creep

High-energy ion-implantation in silicon carbide

A NEW SAMPLE CHANGER FOR IBA ANALYSES

Improvements of the prototype IBA sample changer design

A.M. Müller, M. Lezaic, C. Vockenhuber, scientific and technical staff of Laboratory of Ion Beam Physics

In 2021, a prototype sample changer (Fig. 1) was commission for the routine operation at the RBS and ERDA experiments. The implementation of a load-lock box equipped with a small turbo pump minimized the time needed to exchange sample holders. Therefore, the number of measured samples per day was significantly increased.

Fig. 1: *Prototype sample changer for the use at the RBS and ERDA vacuum chamber.*

The linear and rotational stage together with the capability of x-y movement (±5 mm) allows to position the sample over a wide range.

Initially only one sample changer was built compatible with the ERDA and RBS vacuum chamber. Depending on the measurement schedule one has to move the sample change from one chamber to the other quite frequently.

Therefore, an improved version was designed dedicated for the use at the RBS chamber in the frame of the final thesis of our physics laboratory technician Milan Lezaic.

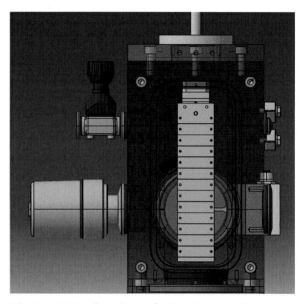

Fig. 2: *CAD drawing of the improved sample changer lock housing. The sample holder comprises in total 24 standard positions (12 on each side).*

The main design change comprises a new lock housing, which facilitates the exchange of the sample holder by a larger top cover (Fig. 2). Additionally, a larger adjustment range (±10 mm) in x-y direction was implemented. The linear stage can now access the entire path from the lock into the chamber, whereas in the prototype version the sample holder had to be transferred by hand.

Due to supply shortages of certain key components the final assembly was delayed by several months. We expect to put the sample changer in operation early 2023.

NOVEL GASES FOR GAS IONIZATION CHAMBERS

Study of heavier gases used in gas ionization chamber detector

M. Kivekäs[1], A.M. Müller

Gas ionization chamber detectors (GIC) are widely used in ion beam analysis (IBA) and in accelerator mass spectrometry (AMS) as energy detectors. GICs have superior resolution, when detecting heavy particles, compared to conventional semiconductor detectors [1]. Nowadays standard gas used in GICs is isobutane, but some others are also used in special cases.

We studied GIC characteristics then some novel gases are used. Gases chosen are hexane and octane, which are hydrocarbons like isobutane, but are heavier molecules. More massive molecule means we require less gas pressure inside the detector to stop incident ions. Lower pressure increases mean free paths of electrons (and ions) in gas, and we anticipated GIC to be able to handle very large count rates better than with isobutane.

First, we measured GIC characteristics using isobutane as a baseline. Detector resolution for protons, ^{12}C and ^{127}I are determined and then again using high count rates. Count rates span from 100 to 30 000 counts/s. After isobutane we switched gas to hexane and repeated measurements. Same with octane.

In our measurements calibrations (Fig. 1) between gases are nearly identical. However, hexane and octane did not increase the detector resolution (peak width) (Fig. 2) like we anticipated. Resolution measured with isobutane and hexane are nearly identical. Hexane seems suitable for GICs and octane also when less resolution is accepted.

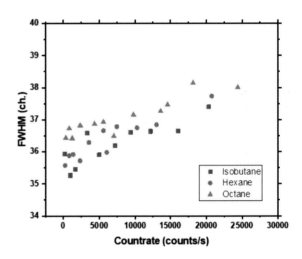

Fig. 2: *Peak widths of 2.5 MeV ^{127}I in different gases in function of count rate. Heavier gases do not result increased resolution with higher count rates.*

There are also advantages using heavier gases. Lower pressure allows thinner beam entrance windows that induces less straggling in the window. Other advantage is that hexane and octane are liquids in air pressure. No need for gas handling, liquids require a reservoir connected to detector and gas is then vaporized by the vapor pressure.

[1]   A. M. Müller et al., NIM-B (2017) 40

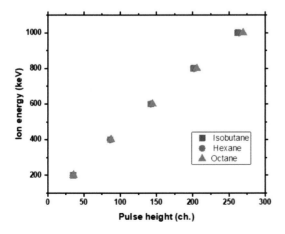

Fig. 1: *Calibration of ^{13}C when measuring with different gases in GIC. Calibrations are nearly identical.*

[1]*Accelerator Laboratory, Univ. of Jyväskylä, Finland*

ELEMENTAL DEPTH PROFILES OF WEATHERED POLYMERS

Assessment of ERDA and FT-IR in weathered polymer films

S. Müller, C. Vockenhuber, A.M. Müller, R. Kaegi[1]

Plastic polymers can be found everywhere in our lives due to their advantages and variety in properties such as their chemical stability and light weight. However, by now it is commonly known that a large part of the millions of tons of plastic that is produced per year ends up in the environment or the sea and decays too slowly to be sustainable. To open a path to solutions we study how and how quickly different polymer films decay under UV radiation. This simulates a common aging process of plastic in nature, the aging under UV radiation from sunlight.

Application of elastic recoil detection analysis (ERDA) gives an insight into the composition of the polymers: The beam is directed at a polymer foil under glancing angle and recoil atoms are detected under the same angle. In He-ERDA the primary He ions are stopped in an absorber foil in front of the detector allowing to measure only H. From the energy distribution the H depth profile is measured and it is possible to characterize how it changes under different radiation times up to 1000 h.

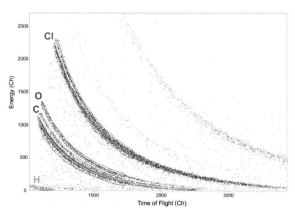

Fig. 1: *Energy vs time-of-flight of the polymer PVC showing a 'mass banana' for each element. The sample was exposed to UV light for 250 h which increased the oxygen content.*

Polymers are mostly composed of hydrogen and carbon (such as PE, PS, PP) but can also include oxygen (PLA) or chlorine (PVC). Exposure to UV radiation will oxidize the samples and increase the amount of oxygen in the composition. The investigation of these heavier elements is conducted by heavy ion ERDA (HI-ERDA) with a 13 MeV ^{127}I beam. The energy vs. time-of-flight spectra show a 'mass banana' (Fig. 1) for each element and the intensities at different energies make it possible to generate elemental depth profiles. We expect to find that oxidation predominantly takes place at the surface. Special care is taken to minimize beam damage during the ERDA measurements and changes of counting rates are considered in the data evaluation.

We also use Fourier transform infrared (FT-IR) spectroscopy to specifically investigate the carbonyl index of the samples. Using IR light to excite different vibrational modes results in a characteristic spectrum for every polymer. The carbonyl peak at wavelengths between 1600-1800 cm^{-1} (Fig. 2), from which the carbonyl index can be calculated, is expected to be higher for oxidized samples.

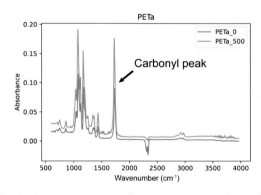

Fig. 2: *FT-IR spectrum for comparison of PET foil at 0 h and 500 h UV exposure time. The carbonyl peak around 1700 cm^{-1} characterizes the oxidation of the sample.*

[1]*Microplastics and analytical methods, Eawag, Dübendorf*

REACTION FRONT CONTROL IN Ni/Al RM

Temperature and speed of reaction control via structural variation

N. Toncich[1], A.M. Müller, C. Vockenhuber, R. Spolenak[1]

Reactive Multilayers (RM) are metastable systems mainly composed of at least two different stacked, well-defined layers (Fig. 1). The main feature of the reactants is the large negative enthalpy and the high adiabatic reaction temperature. The most widely used method for producing reactive multilayers is Physical Vapor Deposition (PVD). This technique offers good control over the architecture of the multilayers and thus allows excellent tuning of the behavior of these systems [1].

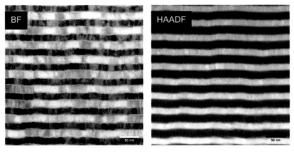

Fig. 1: *Bright-Field (BF) and High-Angle Annular Dark-Field (HAADF) Scanning Transmission Electron Microscopy (STEM) images of 30-fold repetition Ni/Al RM.*

Tailoring the RM behavior would lead to a wider range of applications. Therefore, a good calibration of the deposition parameters and thus of the resulting thickness of the individual layers is extremely important in order to be able to study different structural variations that enable to control the reaction propagation speed and the reached temperatures (Fig. 2).

In the present work, alternated Ni and Al layers were magnetron sputtered with a 4-fold repetition on a Si-wafer. The sputtering parameters were set to achieve a 40 nm thick Ni/Al bilayer.

Fig. 2: *Frames of the reaction front propagation in Ni/Al RM.*

The sample was measured with Rutherford Backscattering Spectroscopy (RBS) with a 2 MeV He beam. The result showed the presence of a 15.5-nm-thick Ni layer and a 25.5 nm thick Al-layer, repeated four times. The result obtained is comparable to what was expected.

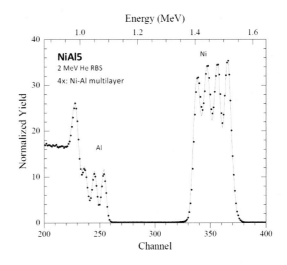

Fig. 3: *RBS measurement of 4-fold repetition Ni/Al RM.*

[1]   D. P. Adams, Thin Solid Films 576 (2015) 98

[1]*Laboratory for Nanometallurgy, ETH Zurich*

LITHICONE COATINGS FOR Li-ION BATTERIES

Li organic-inorganic coating promote cathode capacity performance

K. Egorov[1], W. Zhao[1], K. Knemeyer[2], A.N. Filippin[2], C. Vockenhuber, A. Giraldo[2], C. Battaglia[1]

Nobel laureate Stanley Wittingham and coauthors recently pointed out that the capacity and the energy density of lithium-ion batteries with $LiNi_{0.8}Mn_{0.1}Co_{0.1}O_2$ (NMC811) cathodes may be increased significantly by mitigating the so-called first-cycle capacity loss [1].

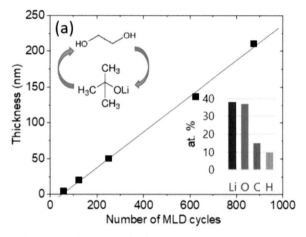

Fig. 1: *Lithicone thickness vs. the number of MLD cycles determined by ellipsometry. The inset shows the lithicone composition as determined by ERDA.*

Here we show that capacity losses associated with loss of lithium due to irreversible reactions with the electrolyte and/or loss of active cathode material due to irreversible structural changes can be mitigated employing an artificial cathode electrolyte interphase (CEI) layer. We employ molecular layer deposition (MLD) to grow so-called lithicone layers (Fig. 1) [2] directly onto porous NMC811 particle electrodes, thereby maintaining the electronic contact between NMC811 particles established by carbon black particles. The Lithicone composition was determined using elastic recoil detection analysis (ERDA) using 13 MeV ^{127}I ions and evaluated with the Potku software [3]. The results show that the lithicone layer consists of 38 at.% Li, 37 at.% O, 15 at.% C, and 10 at.% H resulting in a layer stoichiometry of $LiOC_{0.4}H_{0.3}$. The growth rate

extracted from the slope of the linear fit is about 2.5 Å/cycle.

We demonstrate that lithicone is very efficient in reducing capacity losses during the first cycles resulting in an overall capacity increase of 5% during long-term cycling without negatively affecting the cell's rate capability (Fig. 2)

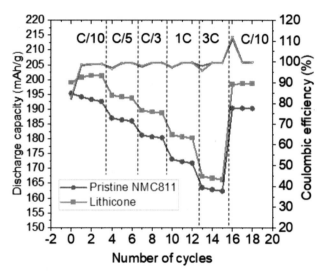

Fig. 2: *Formation cycles and variation of discharge capacity vs. current rates for NM811||graphite coin cells. 1C current was about 0.95 mA.*

[1]   H. Zhou et al., ACS Energy Lett. 4 (2019) 1902

[2]   E. Kazyak et al., Chem. Commun. 56 (2020) 15537

[3]   K. Arstila et al., NIM-B 331 (2014) 34

[1]EMPA, Dübendorf
[2]BASF Schweiz AG, Basel

COMPOSITION OF ANODIC ALUMINUM OXIDES

Influence of anodizing parameters on the growth of anodic Al oxides

N. Ott[1], S. Radeck[1], M. Döbeli, A.M. Müller, P. Schmutz[1]

Anodizing is a process widely used in the industry as a surface finish for aluminum substrates to provide corrosion protection, wear resistance, insulating properties, and decorative appearance. By varying the anodizing conditions (electrolyte, applied voltage or current), the structure of the formed anodic Al oxide can be tuned and subsequently functionalized, making it highly attractive for various industrial applications, ranging from large lightweight structures down to nanoscale electronics, optoelectronics, catalysts, and sensors. With the increased use of aluminum in our societies and the emergence of new manufacturing routes, understanding the anodic oxide growth mechanisms, especially with regards to the intermediate barrier(-like) oxide layer, is essential to ensure homogeneous current flow during anodizing of high-aspect ratio, structured surfaces, or geometrically complex components.

Rutherford backscattering spectrometry (RBS) and complementary elastic recoil detection analysis (ERDA) were performed to determine the composition of these intermediate barrier-like anodic Al oxide layers. Figure 1 shows that the two layers have similar Al/O ratio, with an excess of oxygen compared to Al_2O_3. Al anodic oxide layers in fact consist of a complex arrangement of Al oxides, oxihydroxides, and hydroxides with additional incorporation of electrolytic anions, as evidenced by the presence of P and respectively, S in the RBS spectra.

Interestingly, the barrier-like anodic oxide layer formed in phosphoric acid (H_3PO_4) presents a density close to that of a true high-field controlled Al barrier layer. In comparison, the barrier-like Al anodic oxide layer formed in sulfuric acid (H_2SO_4) is highly defective. These findings confirm that the barrier-like Al anodic oxide layer formed in the early stages defines the subsequent growth of porous anodic oxides.

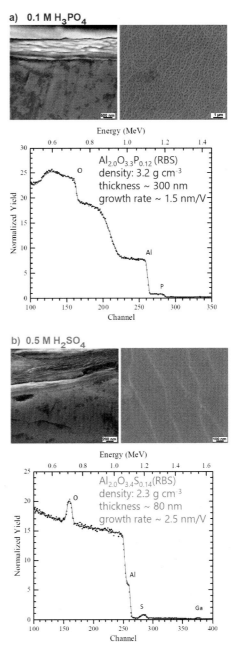

Fig. 1: Intermediate barrier(-like) anodic oxide layers obtained for pure Al (99.5%) anodized in a) 0.1 M H_3PO_4 and b) 0.5 M H_2SO_4.

[1]Laboratory for Joining Technologies and Corrosion, Empa, Dübendorf

ORIGIN OF IRRADIATION CREEP

Trace diffusion studied by in-beam creep testing, ex-situ TEM and SIMS

A. Nastruzzi[1,2], J. Chen[1], W. Jiang[3], M.A. Pouchon[1,2], A.M. Müller, C. Vockenhuber

Materials are one of the key factors for safe and reliable operation of nuclear reactors. The prediction of advanced nuclear material property changes will serve the increasing demand of energy but also requires the development of new materials. These materials have to be able to withstand extreme conditions: high temperatures, corrosive environments and extreme radiation exposures. Here irradiation creep becomes important and the main processes underlining the response of the irradiated material should be recognized. The most important models, proposed to understand the irradiation creep mechanism (SIPA, SIPN and CCG), cannot completely explain the experiment evidences.

In the present study on radiation effects, high-pure FCC Ni single crystal with 20 nm of Ni-64 layer and three different lattice orientations (100, 110, 111) was chosen as representative material.

The three main investigation methods are: 1) in-beam irradiation creep test with 7.2 MeV proton irradiation, at various temperatures and loadings, 2) SIMS and RBS trace diffusion measurement and 3) ex-situ TEM observation.

The objective is to predict the irradiation creep phenomenon with three main points: 1) the effects of loading direction with respect to lattice orientation on irradiation creep compliance (B0), 2) the impacts of stress/strain on the irradiation-enhanced self-diffusion and 3) the influence of applied stress on density and size distribution of dislocation loops.

The combination of SIMS and RBS will allow us to accurate determine the Ni self-diffusion coefficient after the irradiation creep tests. In order to check the RBS resolution, samples 1_Ni-100, 1_Ni-110 & 7_Ni-100 have been measured by 2 MeV He RBS, before irradiation.

An example of RBS spectrum of 1_Ni-110 is reported in Fig. 1.

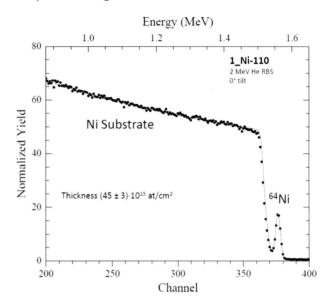

Fig. 1: *Example of RBS spectrum of the sample 1_Ni-110 with 10 nm Ni-64 layer.*

All investigated samples showed a well identifiable Ni-64 signal, so this technique will be used to analyze the Ni self-diffusion after the irradiation in combination with SIMS.

In conclusion, the main goal of this study is to correlate the main processes of the irradiated material response to stress and irradiation, in order to develop a new model in the irradiation creep theory, where diffusion plays an important role.

---

[1]*Paul Scherrer Institute, Villigen*
[2]*Materials Science and Engineering, EPFL, Lausanne*
[3]*Key Laboratory of Materials Physics, Chinese Academy of Sciences, Hefei, China*

HIGH-ENERGY ION-IMPLANTATION IN SILICON CARBIDE

Deep implantation in SiC with high-energy ions for high-power devices

M. Belanche[1], C. Martinella[1], A.M. Müller, U. Grossner[1]

Due to its superior physical properties, Silicon carbide (SiC) has received much attention as viable alternative to silicon for high-power and high-temperature devices. Unlike in silicon, diffusion doping is not feasible in SiC and selective doping for device applications is exclusively done by ion implantation. Throughout the last two decades, significant effort was made to improve fundamental understanding of the implantation process to reduce performance- and reliability-limiting issues such as crystal damage [1], surface roughening [2], lateral straggling [3] or channeling [4]. The species used in commercially available SiC power devices are Nitrogen, Aluminum or Phosphorous, implanted at relatively low energies ranging from hundreds of keV up to a few MeV, which results in the ions stopping within the first three micrometers of the SiC crystal for the highest energies. However, future SiC device designs will require implantations at much deeper depths requiring ten times higher implantation energies (Fig. 1). Hence, a better understanding of damage creation, channeling effects and post-irradiation annealing schemes in such regimes is needed before their use in the fabrication of high-power devices.

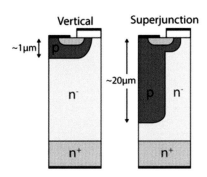

Fig. 1: *Left: Vertical MOSFET configuration (commercially available). Right: Super junction MOSFET configuration (future device with improved performance). P-type implanted region in red.*

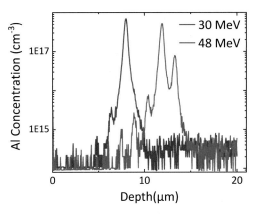

Fig. 2: *SIMS analysis of the depth profile of implanted Al[4+] and Al[7+], at 30 MeV and 48 MeV, respectively.*

Thanks to the collaboration with the Laboratory of Ion Beam Physics (LIP) and by using their available 6 MV Tandem accelerator, we are able to study the effect on SiC of high-energy ion implantation up to 50 MeV with expected penetrations depths up to 20 µm in the crystal, and with different ion doses and species. After irradiation, the SiC samples are characterized with techniques such as SIMS (Fig. 2), to study effects such as penetration depths and energy- and dose-dependent damage creation. The goal is to create an experimental base combining different ion energies, doses and species at the regimes aforementioned and their effects on the SiC crystal.

[1]   R. Nipoti et al., Mater. Sci. Semicond. Process 78 (2018) 13
[2]   M. A. Capano et al., J. Electron. Mater. 27 (1998) 370
[3]   J. Müting et al., Mater Appl. Phys. Lett. 116(1) (2020)
[4]   M. K. Linnarsson et al., Semicond. Sci. Technol. 34 (2019)

[1]Advanced Power Semiconductor Laboratory, ETH Zurich

EDUCATION AND OUTREACH

Attendees of the 24th Radiocarbon Conference and the 10th 14C & Archaeology Conference at ETH Zurich Polyterrasse on a sunny September 13th

24th Radiocarbon Conference in Zurich

Workshop on compound-specific 14C analysis

Workshop on radiocarbon in ocean water

Local geologic tours

Die Nacht der Physik

Evidence and experiment - Science for public

Education of physics laboratory assistants

24ᵗʰ RADIOCARBON CONFERENCE IN ZURICH

A scientific highlight after a period of online conferences

The local conference organizing team

Fig. 1: *The official conference photo taken in front of the ETH Zurich main building.*

The 24ᵗʰ Radiocarbon Conference was held together with the 10ᵗʰ Radiocarbon and Archaeology Symposium at the ETH main building in Zurich, September 11-16, 2022 (Fig. 1). After a long period of online meetings, the event could take place in person, and we all enjoyed extensive scientific discussions which we all missed very much during the past years. 400 participants from 37 different countries have registered (Fig. 2). There was the option of online participation, and about 40 people took the opportunity to follow the streams of the conference presentations. To accommodate the different time zones around the globe, streams were recorded and made available throughout the conference. Prior to the official conference dates workshops were organized to highlight specific topics. The number of participants who were taking advantage of this may indicate that such more informal platforms for discussions are really needed. Within the five major directions: Archaeology, Technical Developments, Cosmogenic Nuclides, Climate & Calibration, Global Carbon Cycle research, and Tracers more than 420 abstracts have been submitted. To provide the participants with a best suited environment for presenting their scientific results, the program committee decided to organize three parallel oral sessions as well as

three dedicated poster sessions. Facility reports were allocated to a special poster session which was on display throughout the conference. With parallel oral sessions it is not always easy to follow the most interesting presentation, but having all presentations recorded, participants could revisit individual talks. Indeed, many took advantage of this opportunity.

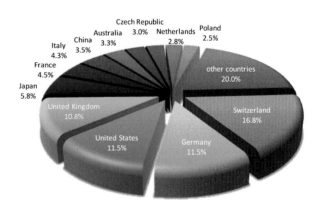

Fig. 2: *Breakdown of conference participants versus country of origin.*

The conference program was completed by an interesting social program. There was the opportunity to visit the company Ionplus, to join sightseeing tours in Zurich, or to exchange with colleagues in an informal atmosphere at the aperitifs and the conference dinner.

WORKSHOP ON COMPOUND-SPECIFIC ^{14}C ANALYSIS

State-of-the-art methods and discussions on future directions

C. Welte, T.I. Eglinton[1], E. Casanova[2], T. Knowles[3]

A series of workshops accompanied the 24th Radiocarbon conference that was held in Zurich in September 2022. One of the workshops focused on compound-specific radiocarbon analysis (CSRA) [1]. The goals of the workshop were two-fold: to bring the community together for an informal exchange on CSRA methods, and to invite those interested in applying this analytical approach for discussion to generate an impression of the opportunities and challenges in this research area. In the first part of the workshop, the focus was on sharing methods used for purification of specific compounds. An overview of state-of-the-art instruments (see Fig. 1), suitable approaches for blank assessment [2] and strategies for the isolation of different compounds was given in a series of six presentations. A coffee break allowed for an informal exchange between the participants, networking and exploring opportunities for future collaboration.

Fig. 1: *Preparative gas chromatograph in the 1950s illustrating the origins of this technique that is one of the main separation techniques for CSRA today (sketch by Tim I. Eglinton).*

In the second part, a status report regarding an on-going CSRA intercomparison exercise using well-characterized "bog butter" reference material, and organized by the BRAMS facility at Bristol University, UK, was presented.

Fig. 2: *Conveners of the workshop (from left to right: Tim I. Eglinton, Tim Knowles, Emmanuelle Casanova, Caroline Welte).*

This formed the basis for subsequent discussions in break-out sessions on four topics that were moderated by experts: 1. Intercomparison exercises for other compounds (G. Mollenhauer), 2. Foreseen technical developments (R. Smittenberg), 3. Other compounds requiring CSRA (N. Ishikawa), 4. New applications of CSRA to other fields (L. Hendriks). The workshop concluded with each moderator presenting the main findings of their discussion in a plenary session. The majority of the 53 participants supported the notion that CSRA community should meet on a regular basis and expressed a desire for establishing a mailing list. After the workshop, a compilation of the methods presentations and summary of the discussions was distributed among the participants.

[1]   T. I. Eglinton et al., Anal. Chem. 68 (1996) 904

[2]   N. Haghipour et al., Anal. Chem. 91 (2019) 2042

[1]*Earth Sciences, ETH Zurich*
[2]*BRAMS, School of Chemistry, University of Bristol, UK*
[3]*Muséum National d'Histoire Naturelle, Paris, France*

WORKSHOP ON RADIOCARBON IN OCEAN WATER

^{14}C in dissolved inorganic carbon of seawater

M. Castrillejo[1], L. Wacker, L. Raimondi[2], K. Kündig[2]

During the 24th Radiocarbon International Conference, Zurich, we organized the workshop 'Radiocarbon in ocean water'. The aim was to provide first-hand experience in the laboratory with seawater sample processing and to discuss different methodologies for the determination of radiocarbon in dissolved inorganic carbon (DIC) of seawater.

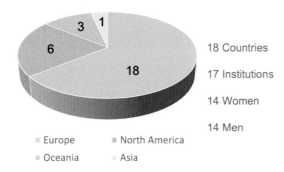

18 Countries

17 Institutions

14 Women

14 Men

Fig. 1: *Statistics of the workshop participants.*

A total of 28 participants (Fig. 1), 14 women and 14 men signed to the workshop. Most attendants came from European countries. One third travelled from more distant destinations in North America, Oceania, and Asia. The organizers were also well gender balanced. Overall, the workshop gathered a diverse group of attendants, including academics and companies interested in marine radiocarbon measurements. Most participants had prior experience with Accelerator Mass Spectrometry (AMS) measurements of radiocarbon. However, very few of them had the capabilities for processing seawater samples and were highly interested in the new simplified method allowing the rapid analysis of radiocarbon in DIC of seawater developed at the Laboratory of Ion Beam Physics [1].

The workshop consisted of a welcome and introductory part, a lab tour, and short presentations. Participants were arranged in three smaller groups and guided through the radiocarbon laboratory and the AMS measurement hall. By using real samples and data (Fig. 2), they learnt about sample collection and handling, the extraction and graphitization steps, and the AMS measurement and data evaluation.

Fig. 2: *Seawater samples ready for carbon extraction during the laboratory tour.*

After the lab tour, we smoothly transitioned to discuss other methods and the analytical challenges that the radiocarbon community is facing. The discussion begun with 4 short presentations given by scientists from the Woods Hole Oceanographic Institution (WHOI), the Alfred Wegener Institute (AWI), the Australian Nuclear Science and Technology Organisation (ANSTO), and ETH-Zurich. The speakers talked about the WSL and REDICS lines from WHOI, the performance of marine measurements using gas interface AMS at AWI, and the building of a new line for seawater at ANSTO.

[1]   N. Casacuberta et al., Radiocarbon 62(1) (2020) 13

[1]*Space & Atmospheric Physics, Imperial College London, UK*
[2]*Physical Oceanography, ETH Zurich*

LOCAL GEOLOGIC TOURS

Quaternary Geology of Uetliberg and geological city tour Zurich

S. Kamleitner, S. Ivy-Ochs, E. Broś, A. S. Søndergaard

Within the scope of the 24th Radiocarbon and the 10th 14C & Archeology International conferences hosted in September of 2022, two walking tours on the local geological history and highlights of Zurich were organized by LIP.

The guided excursion to Uetliberg, offered intresting insights into the geology and geomorphology of Zurich's 'Hausberg'. The 35 participants enjoyed the spectacular view down onto the city and Lake Zurich (Fig. 1). Next they visited the famous Leiterli outcrop of Early Pleistocene glaciofluvial gravels ('Decken-schotter') located just south of Uto Kulm. A visit to Fallätscha provided insight into the Miocene Freshwater Molasse sandstones. Middle Pleistocene and Last Glacial Maximum glacial landforms, such as the lateral moraines along Lake Zurich, allowed to envision how the region must have looked like during the peak of the last glaciation.

Fig. 1: *Participants on the Uetliberg excursion enjoying the view of the city of Zurich and Lake Zurich (credit: M. Bolandini).*

Geological processes and materials shaping Zurich's cityscape were explored during the geological city tour. The excursion led 14 participants from ETH main building via Neumarkt, Rathaus, Grossmünster and Frauen-münster, to the frontal moraine ridge and former fort of Lindenhof (Fig. 2). Various building stones of different age and originating from across Switzerland could be seen along the way and allowed to inspect characteristics (colour, texture, fossils) of different rock types. The role of ice marginal deposits related to the last glaciation and the impact of (natural and artificial) lake level variations and river course changes on the appearance of Zurich city were discussed alongside. Accompanied by mystic legends and historic incidents, participants further gained a brief overview on Zurich's settlement history, from Roman and Medieval into modern times.

Fig. 2: *Participants of the Geological City Tour at Lindenhof, view towards the old town and Grossmünster (credit: A. Sironić).*

Alternatively to the geological excursions, conference participants were invited to join guided tours on the city's water management and old town or paid a visit to different archaeological sites, the National Museum, the Kunsthaus Zurich, the Semper Observatory or the Botanical garden.

DIE NACHT DER PHYSIK

Open house to visit scientific activities at ETH Zurich

H.-A. Synal, scientific and technical staff of Laboratory of Ion Beam Physics

Fig. 1: *View of the high energy end of the 6 MV Tandem accelerator inside the HPK building.*

On the afternoon and evening of Friday, June 17, 2022, the Department of Physics was pleased to invite the general public interested in science to a "Night of Physics" at the ETH campus Hönggerberg. An exciting and entertaining program mix was offered for young and old, which also left enough room for direct and spontaneous exchange with researchers, students, and to explore the experimental infrastructure on campus. The Laboratory of Ion Beam Physics, hosting two particle accelerators capable of providing beams of heavy ions between a few hundred keV and 50 MeV, and five dedicated Accelerator Mass Spectrometer (AMS) instruments opened their doors. Many visitors took the opportunity to get an insight view into such an elaborate instrument park that enables modern cutting-edge research.

During 20 min laboratory tours, visitor groups of 5 - 8 people experienced the instruments which are used to detect individual atoms. Of course, radiocarbon dating was the most fascinating topic and people were pleased to experience how from tiny amounts of organic materials, e.g., from the tissue of the "Ice Man", the age of origin of that material can be derived. The instruments used for this purpose are accelerator systems that were originally developed for basic nuclear physics research. The historic development

process of the applied technologies could be followed, demonstrating the progress of the instruments.

Fig. 1: *Dr. Arnold Müller is explaining the secrets of radiocarbon dating to a group of visitors.*

Beside the laboratory tour, a special lecture was offered highlighting the principles and opportunities of state-of-the-art radiocarbon dating which goes well beyond the traditional archaeological applications of the method. Communicating physics in an entertaining way is the main objective of events such as the "Night of Physics" and the fascinating challenges related to dating historic objects were well received by our visitors.

EVIDENCE AND EXPERIMENT - SCIENCE FOR PUBLIC

The Anthropocene research and the event at the museum

I. Hajdas, K. Wyss

The debates surrounding the Anthropocene intensify as the proposal for the definition of this new chronostratigraphic unit is prepared [1]. The research completed by the Anthropocene Working Group (AWG) was presented and discussed during a meeting held in May 2022 at the Haus der Kulturen der Welt (HKW), a large cultural institution in the center of Berlin (Fig. 1).

Fig. 2: *Preparation of samples for the AWG/HKW project. Sediment samples from the Baltic Sea are transferred into reaction tubes for ABA treatment.*

Fig. 1: *The scientific meeting of the AWG was held at the HKW in Berlin in 18-22 May 2022. The meeting was accompanied by an art/science exhibition and public discussions.*

The HKW together with the Max Planck Institute for the History of Science have accompanied the scientific debate associated with the Anthropocene and the work of the AWG since 2013. A documentation about this pioneering collaboration between the sciences, the humanities and the arts can be found at www.anthropocene-curriculum.org, including a wealth of materials on the AWG's comprehensive research and recordings of the scientific event in May 2022.

Documentation of the ^{14}C sample preparation (Fig. 2) and annotated pictures of the laboratory were part of the exhibition by the artists Giulia Bruno and Armin Linke (Fig. 3).

Fig. 3: *An impression from the exposition at the HKW showing the documentation of the AMS ^{14}C analysis at ETH Zurich. The arts exhibition was created by the artists Giulia Bruno and Armin Linke.*

[1]   C. Waters and S. Turner, Science 378 (2022) 706

EDUCATION OF PHYSICS LABORATORY ASSISTANTS

New sample combusting system (SCS)

L. Wacker, M. Huwyler, P. Gautschi, S. Bühlmann

The Laboratory of Ion Beam Physics (LIP) takes part in the educational program for apprentices of physics laboratory assistants. The students join our group and learn how to engineer scientific equipment.

Starting in 2020, Severin Zimmermann and later Damian Sonderegger built for their final exams a new sample combustion system (SCS), that will allow for more efficient sample combustion with minimal risk of sample contamination. The SCS will also better fit to the existing graphitization equipment and be more cost efficient compared to the presently used elemental analyser (EA). Marcel Huwyler now completed the SCS by adding a gas handling system in his final exams after he already developed a sample introduction system for the SCS last year [1].

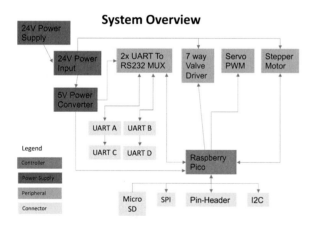

Fig. 1: Electronic board specifically designed to control the components of the SCS.

A new electronic board was designed for a Raspberry Pi microcontroller (Fig. 1), that allows to control all components via serial communication and digital I/Os. Major components to control are the mass flow controllers for helium carrier gas and oxygen (RS232), a pressure transducer (I^2C), a ball valve (Servo PW), sample changer (Stepper motor, RS485) and valves of the gas handling (Digital

out). A small python program controls the combust single samples (actions in fig. 2) from a computer connected to Raspberry Pi Pico microcontroller board of the SCS.

The combustion system is constantly flushed with a helium flow of 150 ml/min. The IR signal for the CO_2 measurement at the end of the line is given in fig. 2 for an organic sample containing about 1 mg of carbon.

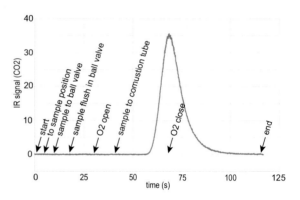

Fig. 2: The blue curve shows the IR signal over time. Indicated are the actions of the combustion sequence. The combustion is completed in less than 120 s.

The system does not separate the combustion gases nitrogen and carbon dioxide as the presently used elemental analysers from *Elementar* for the sample combustion because it is not required for a successful graphitisation with the AGE graphitisation system. The advantage of the new SCS is a reduced time of less than 2 min for the combustion of a single sample in contrast to the 10 min required by the presently used elemental analysers. Additionally, the new system is cheaper and should also be more reliable because the complexity is reduced.

The SCS will next be coupled to a graphitisation system.

[1] L. Wacker et al., LIP annual report (2021) 105

PUBLICATIONS

N. Akçar
Chapter 20 - The Anatolian Peninsula
European Glacial Landscapes. Maximum Extent of Glaciations (2022) 149 - 157

N. Akçar
Chapter 63 - The Anatolian Mountains: glacial landforms from the Last Glacial Maximum
European Glacial Landscapes. Maximum Extent of Glaciations (2022) 497 - 504

N. Akçar, S. Ivy-Ochs, F. Schlunegger
A Special Issue of Geosciences: Cutting Edge Earth Sciences—Three Decades of Cosmogenic Nuclides
Geosciences **12** (2022) 409

J.L. Andersen, A. Margreth, O. Fredin, H. Linge, B.W. Goodfellowf, J.C. Faust, J. Knies, T. Solbakk,
E.J. Brook, T. Scheiber, R.v.d. Lelij, V. Burki, L. Rubensdotter, T. Himmler, S. Yeşilyurt, M. Christl,
C. Vockenhuber, N. Akçar
Rapid post-glacial bedrock weathering in coastal Norway
Geomorphology **397** (2022) 108003

V. Andreaki, J. Barceló, F. Antolín, P. Gassmann, I. Hajdas, O. López-Bultó, H. Martínez-Grau,
N. Morera, A. Palomo, R. Pique, J. Revelles, R. Rosillo, X. Terradas
Absolute chronology at the waterlogged site of La Draga (Lake Banyoles, Ne Iberia): Bayesian chronological models integrating tree-ring measurement, radiocarbon dates and micro-stratigraphical data
Radiocarbon **64** (2022) 907 - 948

F. Antolín, H. Martínez-Grau, B. Steiner, F. Follmann, R. Soteras, S. Häberle, G. Prats, M. Schäfer,
M. Mainberger, I. Hajdas, D. Banchieri
Neolithic occupations (c. 5200-3400 cal BC) at Isolino Virginia (Lake Varese, Italy) and the onset of the pile-dwelling phenomenon around the Alps
Journal of Archaeological Science: Reports **42** (2022) 103375

A. Aribia, J. Sastre, X. Chen, M.H. Futscher, M. Rumpel, A. Priebe, M. Döbeli, N. Osenciat, A.N. Tiwari,
Y. Romanyuk
Unlocking Stable Multi-Electron Cycling in NMC811 Thin-Films between 1.5–4.7 V
Advanced Energy Materials **12** (2022) 2201750

B. Ausín, N. Haghipour, E. Bruni, T.I. Eglinton
The influence of lateral transport on sedimentary alkenone paleoproxy signals
Biogeosciences **19** (2022) 613 - 627

B. Biswas, V.F. Michel, Ø.S. Fjellvåg, G. Bimashofer, M. Döbeli, M. Jambor, L. Keller, E. Müller,
V. Ukleev, E.V. Pomjakushina, D. Singh, U. Stuhr, C.A.F. Vaz, T. Lippert, C.W. Schneider
Role of Dy on the magnetic properties of orthorhombic $DyFeO_3$
Physical Review Materials **6** (2022) 074401

M. Bodik, C. Walker, M. Demydenko, T. Michlmayr, T. Bähler, U. Ramsperger, A.-K. Thamm, S. Tear, A. Pratt, M. El-Gomati, D. Pescia
The effect of work function during electron spectroscopy measurements in Scanning Field-Emission Microscopy
Ultramicroscopy **238** (2022) 113547

N. Brehm, M. Christl, T.D.J. Knowles, E. Casanova, R.P. Evershed, F. Adolphi, R. Muscheler, H.-A. Synal, F. Mekhaldi, C.I. Paleari, H.-H. Leuschner, A. Bayliss, K. Nicolussi, T. Pichler, C. Schlüchter, C.L. Pearson, M.W. Salzer, P. Fonti, D. Nievergelt, R. Hantemirov, D.M. Brown, I. Usoskin, L. Wacker
Tree-rings reveal two strong solar proton events in 7176 and 5259 BCE
Nature Communications **13** (2022) 1196

L. Bruno, B. Campo, I. Hajdas, W. Hong, A. Amorosi
Timing and mechanisms of sediment accumulation and pedogenesis: Insights from the Po Plain (northern Italy)
Palaeogeography, Palaeoclimatology, Palaeoecology **591** (2022) 110881

T. Calligaro, A. Banas, K. Banas, I.B. Radović, M. Brajković, M. Chiari, A.-M. Forss, I. Hajdas, M. Krmpotić, A. Mazzinghi, E. Menart, K. Mizohata, M. Oinonen, L. Pichon, J. Raisanen, Z. Siketic, Z. Smit, A. Simon
Emerging nuclear methods for historical painting authentication: AMS-^{14}C dating, MeV-SIMS and O-PTIR imaging, global IBA, differential-PIXE and full-field PIXE mapping
Forensic Science International **336** (2022) 111327

N. Casacuberta, J.N. Smith
Nuclear Reprocessing Tracers Illuminate Flow Features and Connectivity Between the Arctic and Subpolar North Atlantic Oceans
Annual Review of Marine Science **15** (2022) 203 - 221

M. Castrillejo, N. Casacuberta, C. Vockenhuber, P. Lherminier
Rapidly Increasing Artificial Iodine Highlights Pathways of Iceland-Scotland Overflow Water and Labrador Sea Water
Frontiers in Marine Science **9** (2022) 897729

E. Chamizo, M. Christl, M. López-Lora, N. Casacuberta, A.-M. Wefing, T. Kenna
The potential of ^{233}U/^{236}U as a water mass tracer in the Arctic Ocean
Journal of Geophysical Research: Oceans **127** (2022) e2021JC017790

E. Chamizo, M. López-Lora, M. Christl
Performance of the 1 MV Accelerator Mass Spectrometry system at the Centro Nacional de Aceleradores for the analysis of ^{233}U at environmental levels
Nuclear Inst. and Methods in Physics Research, B **533** (2022) 81 - 89

A. Chan, T. Sadykov, J. Blochin, I. Hajdas, G. Caspari
The polymorphism and tradition of funerary practices of medieval Turks in light of new findings from Tuva Republic
Plos One **17** (2022) e0274537

J.D. Chaplin, M. Christl, A.B. Cundy, P.E. Warwick, P. Gaca, F. Bochud, P. Froidevaux
Time-Integrated Bioavailability Proxy for Actinides in a Contaminated Estuary
ACS ES&T Water **2** (2022) 1688 - 1696

J.D. Chaplin, M. Christl, A.B. Cundy, P.E. Warwick, D.G. Reading, F. Bochud, P. Froidevaux
Bioavailable actinide fluxes to the Irish Sea from Sellafield-labelled sediments
Water Research **221** (2022) 118838

J.D. Chaplin, M. Christl, M. Straub, F. Bochud, P. Froidevaux
Passive Sampling Tool for Actinides in Spent Nuclear Fuel Pools
ACS Omega **7** (2022) 20053–20058

P. Chytry, G. Souza, R. Debastiani, C. Dos Santos, J. Antoine, A. Banas, K. Banas, L. Calcagnile, M. Chiari, I. Hajdas, M. Molnar, P. Pelicon, N. Pessoa Barradas, G. Quarta, F.S. Romolo, A. Simon, J.F. Dias
The potential of accelerator-based techniques as an analytical tool for forensics: The case of coffee
Forensic Science International **335** (2022) 111281

J.R. Cox, E. Blinman, L. Wacker, M.W. Rowe
Premature oxidation during argon plasma cleaning of water-rich radiocarbon samples
Radiocarbon **64** (2022) 21 - 34

H. Cubizolle, J. Argant, K. Serieyssol, F. Fassion, C. Oberlin, A.-M. Dendievel, Y. Deng-Amiot, C. Beaudouin, I. Hajdas, J.N. Haas
Environmental changes during the Late-Glacial and Early Holocene at the Gourd des Aillères mire in the Monts du Forez Mountains (Massif Central, France)
Quaternary International **636** (2022) 9 - 24

A. Cutmore, B. Ausín, M. Maslin, T.I. Eglinton, D. Hodell, F. Muschitiello, L. Menviel, N. Haghipour, B. Martrat, V. Margari, P.C. Tzedakis
Abrupt intrinsic and extrinsic responses of southwestern Iberian vegetation to millennial-scale variability over the past 28 ka
Journal of Quaternary Science **37** (2022) 420 - 440

C. Daxer, M. Ortler, S.C. Fabbri, M. Hilbe, I. Hajdas, N. Dubois, T. Piechl, C. Hammerl, M. Strasser, J. Moernaut
High-resolution calibration of seismically-induced lacustrine deposits with historical earthquake data in the Eastern Alps (Carinthia, Austria)
Quaternary Science Reviews **284** (2022) 107497

C. Dieleman, M. Christl, C. Vockenhuber, P. Gautschi, N. Akçar
Early Pleistocene complex cut-and-fill sequences in the Alps
Swiss Journal of Geosciences **115** (2022) 1 - 25

C. Dieleman, M. Christl, C. Vockenhuber, P. Gautschi, H.R. Graf, N. Akçar
Age of the Most Extensive Glaciation in the Alps
Geosciences **12** (2022) 39

J. Elkadi, B. Lehmann, G.E. King, O. Steinemann, S. Ivy-Ochs, M. Christl, F. Herman
Quantification of post-glacier bedrock surface erosion in the European Alps using ^{10}Be and optically stimulated luminescence exposure dating
Earth Surface Dynamics **10** (2022) 909 - 928

H. Fernández, J.-L. García, S.U. Nussbaumer, A.J. Geiger, I. Gärtner-Roer, F. Pérez, D. Tikhomirov,
M. Christl, M. Egli
De-icing landsystem model for the Universidad Glacier (34° S) in the Central Andes of Chile during the past~ 660 years
Geomorphology **400** (2022) 108096

B. Fiałkiewicz-Kozieł, E. Łokas, B. Smieja-Król, S. Turner, F.D. Vleeschouwer, M. Woszczyk, K. Marcisz,
M. Gałka, M. Lamentowicz, P. Kołaczek, I. Hajdas, M. Karpińska-Kołaczek, K. Kołtonik, T. Mróz, S. Roberts,
N. Rose, T. Krzykawski, A. Boom, H. Yang
The Sniezka peatland as a candidate for the Global Boundary Stratotype Section and Point for the Anthropocene series
The Anthropocene Review (2022)

H. Gies, M. Lupker, S. Wick, N. Haghipour, B. Buggle, T.I. Eglinton
Discharge-Modulated Soil Organic Carbon Export From Temperate Mountainous Headwater Streams
Journal of Geophysical Research: Biogeosciences **127** (2022) e2021JG006624

G. Monegato, S. Kamleitner, F. Gianotti, S. Martin, C. Scapozza, S. Ivy-Ochs
The Ticino-Toce ice conveyor belts during the Last Glacial Maximum
Alpine and Mediterranean Quaternary **35** (2022) 119 - 134

K.E. Grant, V.V. Galy, N. Haghipour, T.I. Eglinton, L.A. Derry
Persistence of old soil carbon under changing climate: The role of mineral-organic matter interactions
Chemical Geology **587** (2022) 120629

L. Guerra, M.A. Martini, H. Vogel, E.L. Piovano, I. Hajdas, R. Astini, A. De Haller, A. Moscariello,
J.L. Loizeau, D. Ariztegui
Microstratigraphy and palaeoenvironmental implications of a Late Quaternary high-altitude lacustrine record in the subtropical Andes
Sedimentology **69** (2022) 2585–2614

S. Häberle, M. Schäfer, R. Soteras, H. Martínez-Grau, I. Hajdas, S. Jacomet, B. Röder, J. Schibler,
S. van Willigen, F. Antolín
Small Animals, Big Impact? Early Farmers and Pre-and Post-Harvest Pests from the Middle Neolithic Site of Les Bagnoles in the South-East of France (L'Isle-sur-la-Sorgue, Vaucluse, Provence-Alpes-Côte-d'Azur)
Animals **12** (2022) 1511

I. Hajdas
Radiocarbon dating method and the protection of cultural heritage
Global Perspectives on Cultural Property Crime (2022) 121 - 127

I. Hajdas
Georges Bonani (1946–2020) and radiocarbon dating at ETH Zurich
Radiocarbon **64** (2022) 491 - 493

I. Hajdas
The radiocarbon world according to Wally
Radiocarbon **64** (2022) 463 - 466

I. Hajdas, L. Calcagnile, M. Molnár, T. Varga, G. Quarta
The potential of radiocarbon analysis for the detection of art forgeries
Forensic Science International **335** (2022) 111292

K.E. Hansen, J. Giraudeau, A. Limoges, G. Massé, A. Rudra, L. Wacker, H. Sanei, C. Pearce,
M.-S. Seidenkrantz
Characterization of organic matter in marine sediments to estimate age offset of bulk radiocarbon dating
Quaternary Geochronology **67** (2022) 101242

S. Ivy-Ochs, G. Monegato, J.M. Reitner
Chapter 15 - Glacial landscapes of the Alps
European Glacial Landscapes. Maximum Extent of Glaciations (2022) 115 - 121

S. Ivy-Ochs, G. Monegato, J.M. Reitner
Chapter 39 - The Alps: glacial landforms prior to the Last Glacial Maximum
European Glacial Landscapes. Maximum Extent of Glaciations (2022) 283 - 294

S. Ivy-Ochs, G. Monegato, J.M. Reitner
Chapter 58 - The Alps: glacial landforms from the Last Glacial Maximum
European Glacial Landscapes. Maximum Extent of Glaciations (2022) 449 - 460

V. Jomelli, D. Swingedouw, M. Vuille, V. Favier, B. Goehring, J. Shakun, R. Braucher, I. Schimmelpfennig,
L. Menviel, A. Rabatel, L.C.P. Martin, P.-H. Blard, T. Condom, M. Lupker, M. Christl, Z. He, D. Verfaillie,
A. Gorin, G. Aumaître, D.L. Bourlès, K. Keddadouche
In-phase millennial-scale glacier changes in the tropics and North Atlantic regions during the Holocene
Nature Communications **13** (2022) 1419

J. Kaiser, S. Abel, H.W. Arz, A.B. Cundy, O. Dellwig, P. Gaca, G. Gerdts, I. Hajdas, M. Labrenz, J.A. Milton,
M. Moros, S. Primpke, S.L. Roberts, N.L. Rose, S.D. Turner, M. Voss, J.A. Ivar do Sul
The East Gotland Basin (Baltic Sea) as a candidate Global Boundary Stratotype Section and Point for the Anthropocene series
The Anthropocene Review (2022)

S. Kamleitner, S. Ivy-Ochs, G. Monegato, F. Gianotti, N. Akçar, C. Vockenhuber, M. Christl, H.-A. Synal
The Ticino-Toce glacier system (Swiss-Italian Alps) in the framework of the Alpine Last Glacial Maximum
Quaternary Science Reviews **279** (2022) 107400

H. Kerschner, S. Ivy-Ochs, C. Schlüchter
The Moraine at Trins and the Alpine Lateglacial
Landscapes and Landforms of Austria (2022) 341 - 354

K.H. Keskitalo, L. Bröder, D. Jong, N. Zimov, A. Davydova, S. Davydov, T. Tesi, P.J. Mann, N. Haghipour,
T.I. Eglinton, J.E. Vonk
Seasonal variability in particulate organic carbon degradation in the Kolyma River, Siberia
Environmental Research Letters **17** (2022) 034007

C. Lawley, Z.P. Tehrani, A.H. Clark, O.V. Safonova, M. Döbeli, V.N. Strocov, T.J. Schmidt, T. Lippert,
M. Nachtegaal, D. Pergolesi
Protagonists and spectators during photocatalytic solar water splitting with $SrTaO_xN_y$ oxynitride
Journal of Materials Chemistry A **10** (2022) 2374 - 2387

M. Morgenbesser, A. Viernstein, A. Schmid, C. Herzig, M. Kubicek, S. Taibl, G. Bimashofer, J. Stahn, C.A.F. Vaz, M. Döbeli, F. Biautti, J.de Dios Sirvent, M.O. Liedke, M. Butterling, M. Kamiński, M. Tolkiehn, V. Vonk, A. Stierle, A. Wagner, A. Tarancon, A. Limbeck, J. Fleig
Unravelling the Origin of Ultra-Low Conductivity in $SrTiO_3$ Thin Films: Sr Vacancies and Ti on A-Sites Cause Fermi Level Pinning
Advanced Functional Materials **32** (2022) 2202226

N. Mozafari, Ç. Özkaymak, Ö. Sümer, D. Tikhomirov, B. Uzel, S. Yeşilyurt, S. Ivy-Ochs, C. Vockenhuber, H. Sözbilir, N. Akçar
Seismic history of western Anatolia during the last 16 kyr determined by cosmogenic ^{36}Cl dating
Swiss Journal of Geosciences **115** (2022) 5

M. Ndeye, H.-A. Synal, M. Séne
Radiocarbon Concentration Measurements in Tree Leaves near SOCOCIM (Rufisque, Senegal), A Cement Factory
Open Journal of Air Pollution **11** (2022) 1 - 12

K. Nicolussi, M. Le Roy, C. Schlüchter, M. Stoffel, L. Wacker
The glacier advance at the onset of the Little Ice Age in the Alps: New evidence from Mont Miné and Morteratsch glaciers
The Holocene **32** (2022) 624 - 638

J.-P. Niemelä, B. Putz, G. Mata-Osoro, C. Guerra-Nuñez, R.N. Widmer, N. Rohbeck, T.E.J. Edwards, M. Döbeli, K. Maćkosz, A. Szkudlarek, Y. Kuzminykh, X. Maeder, J. Michler, B. Andreaus, I. Utke
Mechanical Properties of Atomic-Layer-Deposited Al_2O_3/Y_2O_3 Nanolaminate Films on Aluminum toward Protective Coatings
ACS Applied Nano Materials **5** (2022) 6285 - 6296

R.F. Ott, D. Scherler, K.W. Wegmann, M.K. D'Arcy, R.J. Pope, S. Ivy-Ochs, M. Christl, C. Vockenhuber, T.M. Rittenour
Paleo-denudation rates suggest variations in runoff drove aggradation during last glacial cycle, Crete, Greece
Earth Surface Processes Landforms (2022) 1–20

C.I. Paleari, F. Mekhaldi, F. Adolphi, M. Christl, C. Vockenhuber, P. Gautschi, J. Beer, N. Brehm, T. Erhardt, H.-A. Synal, L. Wacker, F. Wilhelms, R. Muscheler
Cosmogenic radionuclides reveal an extreme solar particle storm near a solar minimum 9125 years BP
Nature Communications **13** (2022) 214

F. Pawełczyk, I. Hajdas, T. Sadykov, J. Blochin, G. Caspari
Comparing analysis of pretreatment methods of wood and bone materials for the chronology of peripheral burials at Tunnug 1, Tuva Republic, Russia
Radiocarbon **64** (2022) 171 - 186

V.G. Peri, N. Haghipour, M. Christl, C. Terrizzano, A. Kaveh-Firouz, M.F. Leiva, P. Pérez, M. Yamin, H. Barcelona, J.-P. Burg
Quaternary landscape evolution in the Western Argentine Precordillera constrained by ^{10}Be cosmogenic dating
Geomorphology **396** (2022) 107984

D. Püntener, P. Šamonil, D. Tikhomirov, P. Daněk, M. Christl, J. Roleček, M. Egli
Soil erosion rates during the Holocene continuity in a forest-steppe landscape
Earth Surface Processes Landforms (2022) 1-21

G. Quarta, I. Hajdas, M. Molnár, T. Varga, L. Calcagnile, M. D'Elia, A. Molnar, J. Dias, A. Jull
The IAEA Forensics Program: Results of the ams ^{14}C intercomparison exercise on contemporary wines and coffees
Radiocarbon **64** (2022) 1513 - 1524

G. Raab, W. Dollenmeier, D. Tikhomirov, G. Vieira, P. Migoń, M.E. Ketterer, M. Christl, J. Stutz, M. Egli
Contrasting soil dynamics in a formerly glaciated and non-glaciated Mediterranean mountain plateau (Serra da Estrela, Portugal)
Catena **215** (2022) 106314

R. Reber, N. Akçar, D. Tikhomirov, S. Yesilyurt, C. Vockenhuber, V. Yavuz, S. Ivy-Ochs, C. Schlüchter
LGM Glaciations in the Northeastern Anatolian Mountains: new insights
Geosciences **12** (2022) 257

M. Schlomberg, C. Vockenhuber, H.-A. Synal, M. Veicht, I. Mihalcea, D. Schumann
Isobar separation of ^{32}Si from ^{32}S in AMS using a passive absorber
Nuclear Instruments Methods in Physics Research Section B: Beam Interactions with Materials Atoms **533** (2022) 56 - 60

E. Serra, P.G. Valla, R. Delunel, N. Gribenski, M. Christl, N. Akçar
Spatio-temporal variability and controlling factors for postglacial denudation rates in the Dora Baltea catchment (western Italian Alps)
Earth Surface Dynamics **10** (2022) 493 - 512

G.D. Smith, V.J. Chen, A. Holden, N. Haghipour, L. Hendriks
Combined, sequential dye analysis and radiocarbon dating of single ancient textile yarns from a Nazca tunic
Heritage Science **10** (2022) 179

C. Steinhoff, N. Pickarski, T. Litt, I. Hajdas, C. Welte, P. Wurst, D. Kühne, A. Dolf, M. Germer, J. Kallmeyer
New approach to separate and date small spores and pollen from lake sediments in semi-arid climates
Radiocarbon **64** (2022) 1191 - 1207

H.M. Stoll, I. Cacho, E. Gasson, J. Sliwinski, O. Kost, A. Moreno, M. Iglesias, J. Torner, C. Perez-Mejias, N. Haghipour, H. Cheng, R.L. Edwards
Rapid northern hemisphere ice sheet melting during the penultimate deglaciation
Nature Communications **13** (2022) 3819

T.J. Suhrhoff, J. Rickli, M. Christl, E.G. Vologina, V. Pham, M. Belhadj, E.V. Sklyarov, C. Jeandel, D. Vance
Source to sink analysis of weathering fluxes in Lake Baikal and its watershed based on riverine fluxes, elemental lake budgets, REE patterns, and radiogenic (Nd, Sr) and ^{10}Be/^{9}Be isotopes
Geochimica Et Cosmochimica Acta **321** (2022) 133 - 154

M. Suter, L.K. Fifield, S. Maxeiner
The impact of the break-up of molecules in the stripper on the AMS performance
Nuclear Instruments Methods in Physics Research Section B: Beam Interactions with Materials
Atoms **533** (2022) 96 - 103

H.-A. Synal
Accelerator Mass Spectrometry: Ultra-sensitive Detection Technique of Long-lived Radionuclides
Chimia **76** (2022) 45 - 51

A.-K. Thamm, J. Wei, J. Zhou, C. Walker, H. Cabrera, M. Demydenko, D. Pescia, U. Ramsperger, A. Suri,
A. Pratt, S.P. Tear, M.M. El-Gomati
Hallmark of quantum skipping in energy filtered lensless scanning electron microscopy
Applied Physics Letters **120** (2022) 052403

I. Usoskin, S. Solanki, N. Krivova, B. Hofer, G. Kovaltsov, L. Wacker, N. Brehm, B. Kromer
Solar cyclic activity over the last millennium reconstructedfrom annual ^{14}C data (Corrigendum)
Astronomy Astrophysics **664** (2022) C3

N. Vandermaelen, K. Beerten, F. Clapuyt, M. Christl, V. Vanacker
*Constraining the aggradation mode of Pleistocene river deposits based on cosmogenic radionuclide depth
profiles and numerical modelling*
Geochronology **4** (2022) 713 - 730

N. Vandermaelen, V. Vanacker, F. Clapuyt, M. Christl, K. Beerten
*Reconstructing the depositional history of Pleistocene fluvial deposits based on grain size, elemental
geochemistry and in-situ ^{10}Be data*
Geomorphology **402** (2022) 108127

M. Veicht, I. Mihalcea, P. Gautschi, C. Vockenhuber, S. Maxeiner, J.-C. David, S. Chen, D. Schumann
*Radiochemical separation of ^{26}Al and ^{41}Ca from proton-irradiated vanadium targets for cross-section
determination by means of AMS*
Radiochimica Acta **110** (2022) 809 - 816

P. Vinšová, T. Kohler, M. Simpson, I. Hajdas, J. Yde, L. Falteisek, J. Žárský, T. Yuan, V. Tejnecký, F. Mercl,
E. Hood, M. Stibal
The biogeochemical legacy of arctic subglacial sediments exposed by glacier retreat
Global Biogeochemical Cycles **36** (2022) e2021GB007126

A.-M. Wefing, N. Casacuberta, M. Christl, P.A. Dodd
*Water mass composition in Fram Strait determined from the combination of ^{129}I and ^{236}U: Changes
between 2016, 2018, and 2019*
Frontiers in Marine Science **9** (2022) 973507

Y. Wu, X. Dai, S. Xing, M. Luo, M. Christl, H.-A. Synal, S. Hou
Direct search for primordial ^{244}Pu in Bayan Obo bastnaesite
Chinese Chemical Letters **33** (2022) 3522 - 3526

M. Yu, T.I. Eglinton, N. Haghipour, N. Dubois, L. Wacker, H. Zhang, G. Jin, M. Zhao
Persistently high efficiencies of terrestrial organic carbon burial in Chinese marginal sea sediments over the last 200 years
Chemical Geology **606** (2022) 120999

J. Zhang, H. Li, M.G. Wiesner, T.I. Eglinton, N. Haghipour, Z. Jian, J. Chen
Carbon Isotopic Constraints on Basin-Scale Vertical and Lateral Particulate Organic Carbon Dynamics in the Northern South China Sea
Journal of Geophysical Research: Oceans **127** (2022) e2022JC018830

TALKS AND POSTERS

F. Antolin, H. Martínez-Grau, R. Soteras, G. Guidobaldi, M. Jaggi, K. Wyss, S. Bernasconi, I. Hajdas
Combined radiocarbon dating and stable isotope analyses on Neolithic cereal finds identifies turning points in crop dynamics (ca. 6000-3000 BC)
Switzerland, Zurich, 11.-16.09.2022, 24th Radiocarbon - 10th 14C & Archaeology International Conferences

V. Beccari, D. Basso, G. Panieri, A. Almogi-Labin, Y. Makovsky, I. Hajdas, S. Spezzaferri
Significance of micro-and macrofauna from seeps along the Israeli coast (Palmahim Disturbance)
Austria, Vienna, 23.-27.05.2022, EGU General Assembly 2022

M. Bolandini, L. Bröder, D. De Maria, T.I. Eglinton, L. Wacker
A new online ramped oxidation (ORO) system for improved coupled thermal and radiocarbon dissection of complex natural organic matter
Switzerland, Zurich, 11.-16.09.2022, 24th Radiocarbon - 10th 14C & Archaeology International Conferences

N. Brehm, M. Christl, H.-A. Synal, A. Bayliss, K. Nicolussi, C. Pearson, N. Bleicher, D. Brown, L. Wacker
Analysis of solar minima by using radiocarbon in tree-rings
Switzerland, Zurich, 11.-16.09.2022, 24th Radiocarbon - 10th 14C & Archaeology International Conferences

N. Brehm, M. Christl, H.-A. Synal, R. Muscheler, F. Mekhaldi, C. Paleari, A. Bayliss, E. Casanova,
T. Knowles, R. Evershed, K. Nicolussi, T. Pilcher, C. Schlüchter, H. Leuschner, C. Pearson, M. Salzer,
P. Fonti, D. Nivergelt, R. Hantemirov, D. Brown, I. Usoskin, F. Adolphi, L. Wacker
Detection of solar events by using radiocarbon in tree-rings
Switzerland, Zurich, 11.-16.09.2022, 24th Radiocarbon - 10th 14C & Archaeology International Conferences

N. Brehm, M. Christl, H.-A. Synal, R. Muscheler, F. Mekhaldi, C.Paleari, A. Bayliss, E. Casanova,
T. Knowles, R. Evershed, K. Nicolussi, T. Pilcher, C. Schlüchter, H. Leuschner, C. Pearson, M. Salzer,
P. Fonti, D. Nivergelt, R. Hantemirov, D. Brown, I. Usoskin, F. Adolphi, L. Wacker
Detection of solar events by using radiocarbon in tree-rings
Switzerland, Zurich, 19.-22.09.2022, Space Climate 8

E. Broś, S. Ivy-Ochs, R. Grischott, F. Kober, C. Vockenhuber, M. Christl, C. Maden, H.-A. Synal
Isochron-burial dating of the oldest glaciofluvial sediments in the northern Alpine Foreland
Scotland, Edinburgh, 09.-11.06.2022, Nordic Workshop on Cosmogenic Nuclides 2022

E. Broś, S. Ivy-Ochs, R. Grischott, F. Kober, C. Vockenhuber, M. Christl, C. Maden, H.-A. Synal
Isochron-burial dating of the oldest glaciofluvial sediments in the northern Alpine Foreland
Switzerland, Zurich, 11.-16.09.2022, 24th Radiocarbon - 10th 14C & Archaeology International Conferences

E. Broś, S. Ivy-Ochs, R. Grischott, F. Kober, C. Vockenhuber, M. Christl, P. Gautschi, C. Maden,
L. Ylä-Mella, J.D. Jansen, M.F. Knudsen, H.-A. Synal
Age of the oldest Quaternary sediments in the northern Swiss Alpine Foreland
Switzerland, Lausanne, 19.11.2022, 20th Swiss Geoscience Meeting (SGM)

L. Calcagnile, I. Hajdas, M. Molnar, T. Varga, I. Major, M. D'Elia , A.T. Jull, A. Simon, G. Quarta
Application of 14C dating in the routine forensic practice: outcome of the IAEA Coordinated Research Project
Switzerland, Zurich, 11.-16.09.2022, 24th Radiocarbon - 10th 14C & Archaeology International Conferences

M. Caroselli, I. Hajdas, P. Cassitti
The dating of dolomitic mortars with uncertain chronology from Müstair Monastery: sample characterization and combined interpretation of results
Switzerland, Zurich, 11.-16.09.2022, 24th Radiocarbon - 10th 14C & Archaeology International Conferences

N. Casacuberta
Observing the Ocean using Radionuclides as Transient Tracers: an update
Switzerland, Zurich, 16.06.2022, Group meeting Nicolas Gruber

N. Casacuberta
Observing the Ocean using Transient Tracers
Germany, Heidelberg, 07.07.2022, Institute's Research Colloquium

N. Casacuberta
Chasing radioactive pollution to trace ocean circulation
Switzerland, Zurich, 31.10.2022, Inaugural talk ETHZ

N. Casacuberta
Observing the Ocean using Radionuclides as Transient Tracers: The TITANICA project
Switzerland, Zurich, 08.11.2022, Earth Sciences Department Colloquium

N. Casacuberta
Observing the Ocean using Radionuclides as Transient Tracers: an update
Switzerland, Bern, 05.12.2022, Climate and Environmental Physics Seminar

N. Casacuberta, D. Wallace
Modular Research Infrastructures: the MORI project
Switzerland, Lausanne, 02.09.2022, Swiss Polar Day

N. Casacuberta, A.-M. Wefing
Artificial radionuclides as ocean tracers: preliminary results on Arctic Century Cruise
Switzerland, Zurich, 02.11.2022, Swiss Society for Hydrology and Limnology

N. Casacuberta, A.-M. Wefing, M. Christl, J.N. Smith
The Arctic and Subpolar North Atlantic Circulation: new insights from ^{129}I and ^{236}U
Japan, Kanagawa, 07.-09.12.2022, Kinet International Symposium

L. Leist, M. Castrillejo, J.N. Smith, M. Christl, L. Raimondi, N. Casacuberta
Tracing Ocean circulation in the AR7W and OVIDE lines using Artifical Raduîonuclides
Iceland, Hafnarfjörður, 09.-11.05.2022, Arctic and Sub-arctic Ocean Fluxes Meeting

M. Castrillejo
DIC radiocarbon measurements at ETH-LIP
Switzerland, Zurich, 11.-16.09.2022, 24th Radiocarbon - 10th 14C & Archaeology International Conferences

M. Castrillejo, R. Hansman, L. Wacker, J. Lester, H. Graven
Ensuring comparability of radiocarbon measurements in dissolved inorganic carbon of seawater between ETH-Zurich and NOSAMS
Switzerland, Zurich, 11.-16.09.2022, 24th Radiocarbon - 10th 14C & Archaeology International Conferences

M. Castrillejo, L. Wacker, J. Lester, H. Graven
Evolution of radiocarbon in the North Atlantic during 1990s-2020 inferred from in-situ observations and model simulations
Switzerland, Zurich, 11.-16.09.2022, 24th Radiocarbon - 10th [14]C & Archaeology International Conferences

M. Castrillejo, L. Wacker, S. Bollhalder, N. Casacuberta, K. Kündig, L. Leist, G. Scacco, H.-A. Synal, A.-M. Wefing
Update of radiocarbon analyses on dissolved inorganic carbon of seawater at ETH-Zurich
Switzerland, Zurich, 14.09.2022, 24th Radiocarbon - 10th [14]C & Archaeology International Conferences

M. Christl
Why do we need AMS?
Switzerland, Morteratsch, 29.08.2022, Geochronology Summer School

D. Dale, M. Christl, A. Macrander, S. Ólafsdóttir, R. Midag, N. Casacuberta
Using anthropogenic radionuclides to trace ocean circulation around Iceland
Switzerland, Zurich, 22.04.2022, Institute for Biogeochemistry and Pollutant Dynamics PhD Congress 2022

D. Dale, M. Christl, A. Macrander, S. Ólafsdóttir, R. Midag, N. Casacuberta
Using anthropogenic radionuclides to trace ocean circulation around Iceland
Iceland, Hafnarfjörður, 09.-11.05.2022, Arctic and Sub-arctic Ocean Fluxes Meeting

D. De Maria, S. Fahrni, L. Wacker, H.-A. Synal
Double Trap Interface: A novel gas handling system for high throughput AMS analysis
Switzerland, Zurich, 11.-16.09.2022, 24th Radiocarbon - 10th [14]C & Archaeology International Conferences

A. Donner, P. Töchterle, C. Spötl, I. Hajdas, X. Li, R.L. Edwards, G.E. Moseley
Combined ^{14}C and [230]Th/U dating of fine-grained cryogenic cave carbonates from a permafrost cave in Greenland
Austria, Vienna, 23.-27.05.2022, EGU General Assembly 2022

M. Duborgel, L.I. Minich, N. Haghipour, B. González-Domíngez, S. Abiven, T.I. Eglinton, F. Hagedorn
Radiocarbon based turnover rates of soil organic matter fractions along climatic and biogeochemical gradients across in Switzerland
Austria, Vienna, 23.-27.05.2022, EGU General Assembly 2022

K. Fenclová, T. Prášek, M. Němec, M. Christl, P. Gautschi
Fluoride target materials for [239]Pu measurement with MILEA AMS
Czech Republic, Mariánské Lázně, 15.-20.05.2022, 19th Radiochemical conference (RadChem)

K. Fenclová, T. Prášek, M. Němec, M. Christl, P. Gautschi, C. Vockenhuber
Preliminary tests of [26]Al fluoride target matrix on AMS system
Austria, Vienna, 26.04.2022, VERA seminar

P. Gautschi, N. Brehm, M. Wertnik, H.-A. Synal, L. Wacker
Radiocarbon analysis of annually published journals
Switzerland, Zurich, 11.-16.09.2022, 24th Radiocarbon - 10th [14]C & Archaeology International Conferences

P. Gautschi, L. Wacker, H.-A. Synal
Direct graphitization from atmospheric air
Switzerland, Zurich, 11.-16.09.2022, 24th Radiocarbon - 10th 14C & Archaeology International Conferences

Y. Gu, H. Lu, I. Hajdas, N. Haghipour, H. Zhang, J. Wu, K. Shao
Radiocarbon dating of small snail shells in loess-paleosol sequence at Mangshan, central China
Switzerland, Zurich, 11.-16.09.2022, 24th Radiocarbon - 10th 14C & Archaeology International Conferences

I. Hajdas
Bomb peak 14C and the proposal for a new Anthropocene epoch
Austria, Vienna, 24.05.2022, VERA Seminar

I. Hajdas
Radiocarbon dating and its applications to cultural heritage
Austria, Vienna, 13.-16.06.2022, IAEA Workshop on Innovative Accelerator Science and Technology Approaches to Sustainable Heritage Management

I. Hajdas
Radiocarbon dating: applications and calibration
Switzerland, Morteratsch, 29.08.2022, Geochronology Summer School

I. Hajdas
Looking back at 20 years as a member of the EGU
Austria, Innsbruck, 10.11.2022, Institute of Geology, University of Innsbruck

I. Hajdas
14C laboratories and what they can do for ECRs
Switzerland, Bern, 15.11.2022, PAGES Early-Career Network

I. Hajdas
The last 50 thousand years-- how radiocarbon analysis can help to unravel events of the past and present
Poland, Lublin, 08.12.2022, Climate change and environment-challenges and inspirations for science

I. Hajdas, F. Caruso, K. Wyss
AMS 14C dating of artifacts - prospects and challenges
Italy, Lecce, 22.-24.06.2022, IAEA Workshop on Applications of Accelerator-Based and complementary Techniques for Forenisc Science

I. Hajdas, G. Guidobaldi, N. Haghipour, K. Wyss
Samples screening and treatment for accurate radiocarbon dating
Switzerland, Zurich, 11.-16.09.2022, 24th Radiocarbon - 10th 14C & Archaeology International Conferences

D. Michalska, I. Hajdas
Grain fractions versus time intervals – mortars radiocarbon dating
Switzerland, Zurich, 11.-16.09.2022, 24th Radiocarbon - 10th 14C & Archaeology International Conferences

S. Sá, L. Hendriks, I. Hajdas, I. Pombo Cardoso
The application of radiocarbon dating of lead white in the study of polychrome stone sculptures
Switzerland, Zurich, 11.-16.09.2022, 24th Radiocarbon - 10th 14C & Archaeology International Conferences

F. Fiorillo, L. Hendriks, I. Hajdas, E. Huysecom
Integrated methodology for the investigation of paintings – The rediscovery of Jan Ruyscher
Switzerland, Zurich, 11.-16.09.2022, 24th Radiocarbon - 10th [14]C & Archaeology International Conferences

C. Heusser, L. Wacker, T.I. Eglinton, C. Welte
Microsublimation as Final Purification Step for [14]C Analysis of Specific Compounds after Chromatographic Separation
Switzerland, Zurich, 11.-16.09.2022, 24th Radiocarbon - 10th [14]C & Archaeology International Conferences

A.T. Jull, I. Hajdas
Radiocarbon Laboratories and the protection of Cultural Heritage–update and discussion
Switzerland, Zurich, 11.-16.09.2022, 24th Radiocarbon - 10th [14]C & Archaeology International Conferences

S. Kamleitner, S. Ivy-Ochs, L. Manatschal, N. Akçar, M. Christl, C. Vockenhuber, I. Hajdas, H.-A. Synal
Last Glacial Maximum glacier fluctuations of the Rhine and Reuss glacier systems
Switzerland, Lausanne, 19.11.2022, 20th Swiss Geoscience Meeting (SGM)

S. Kamleitner, S. Ivy-Ochs, B. Salcher, J.M. Reitner
Ice flow pattern of Late LGM Rhine glacier reconstructed from a new inventory of streamlined subglacial bedforms
Germany, Potsdam, 25.-29.09.2022, Deutsche Quartärvereinigung (DEUQUA)-Tagung 2022

N. Kappelt, R. Muscheler, G. Raisbeck, J. Beer, M. Christl, C. Vockenhuber, M. Baroni, E. Wolff
[36]Cl as a dating tool for deep ice
Iceland, Reykjavík, 11.-13.05.2022, 35th Nordic Geological Winter Meeting 2022

B. Kromer, L. Wacker, M. Friedrich, S. Lindauer, R. Friedrich, K. Treydte., P. Fonti , E. Martinez
Origin and age of carbon in cellulose of mid-latitude tree rings
Switzerland, Zurich, 11.-16.09.2022, 24th Radiocarbon - 10th [14]C & Archaeology International Conferences

B. Kromer, L. Wacker, M. Friedrich, S. Lindauer, R. Friedrich, K. Treydte., P. Fonti , E. Martinez
Origin and age of carbon in cellulose of mid-latitude tree rings
United Kingdom, Silverstone, 16.09.2022, Radiocarbon in the Anthropocene

U. Leuzinger, I. Hajdas, G. Guidobaldi, K. Wyss, W. Imhof
Alpine archaeology and radiocarbon analysis: a match made in heaven!
Switzerland, Zurich, 11.-16.09.2022, 24th Radiocarbon - 10th [14]C & Archaeology International Conferences

H.-T. Lin, J.-M. Chen, Y.-C. Chou, H.-W. Chiang, M. Christl, C.-C. Shen
Ocean tracer [236]U/[238]U research using a new MC-ICPMS method
USA, Hawaii, 09.-14.07.2022, Goldschmidt 2022

D. Michczyńska, M. Jędrzejowski , M. Kłusek, A. Michczyński, F. Pawełczyk, N. Piotrowska, K. Wyss, I. Hajdas
Radiocarbon dating of fossil wood - verification of the effectiveness of various preparation methods
Switzerland, Zurich, 11.-16.09.2022, 24th Radiocarbon - 10th [14]C & Archaeology International Conferences

A.M. Müller, J. Bourquin, A. Herrmann, R. Pfenninger, M. Suter, H.-A. Synal, L. Wacker
The Ionplus AG – An example of commercializing new innovations in ion beam technologies
Germany, Berlin, 05.-07.09.2022, German Conference for Research with Synchrotron Radiation, Neutrons and Ion Beams at Large Facilities (SNI) 2022

A.M. Müller, B. Jenčič, C. Vockenhuber, T. Wulff, H.-A. Synal, A. Herrmann, S. Maxeiner, L. Wacker, M. Christl, P. Gautschi, S. Fahrni, J. Bourquin
Developments of the ETH/Ionplus Cs-Sputter Ion Source towards high Brightness
Rumania, Sibiu, 17.-23.07.2022, 14th European Conference on Accelerators in Applied Research and Technology (ECAART14)

A.M. Müller, M. Kivekäs, M. Döbeli, H.-A. Synal, C. Vockenhuber, U. Ramsperger
Latest developments of gas ionization chambers for AMS and IBA applications
Austria, Vienna, 14.06.2022, VERA Seminar

K. Nakajima, C. Heusser, C. Welte, L. Wacker, T.I. Eglinton
Going with the flow: rapid pollen sorting for radiocarbon analysis
Switzerland, Zurich, 11.-16.09.2022, 24th Radiocarbon - 10th ^{14}C & Archaeology International Conferences

L. Nguyen, A. Nilsson, C. Paleari, S. Müller, M. Christl, F. Mekhaldi, P. Gautschi, R. Mulvaney, J. Rix, R. Muscheler
A new continuous ^{10}Be record for the last 5000 years measured on ice chips from a borehole in East Antarctica
Austria, Vienna, 23.-27.05.2022, EGU General Assembly 2022

L. Nguyen, A. Nilsson, C. Paleari, S. Müller, M. Christl, F. Mekhaldi, P. Gautschi, R. Mulvaney, J. Rix, R. Muscheler
A new continuous ^{10}Be record for the last 5000 years measured on ice chips from a borehole in East Antarctica
Switzerland, Crans-Montana, 02-07.10.2022, International Partnership in Ice Core Sciences - 3rd Open Science Conference

R. Ott, D. Scherler, K. Wegmann, M. D'Arcy, S. Ivy-Ochs, M. Christl, C. Vockenhuber
Decoupling between fluvial aggradation-incision dynamics and paleo-denudation rates during the last glacial cycle, Crete, Greece
Austria, Vienna, 23.-27.05.2022, EGU General Assembly 2022

F. Pawełczyk, Y. Gu, N. Piotrowska, A. Ustrzycka, I. Hajdas
^{14}C dating for MODIS 2 carbonate mortars – do time and size matter?
Switzerland, Zurich, 11.-16.09.2022, 24th Radiocarbon - 10th ^{14}C & Archaeology International Conferences

P. Povinec, I. Kontuľ, A. Cherkinsky, I. Hajdas, Y. Gu, A.T. Jull, T. Lupták, M. Molnar, P. Steier, I. Svetlik
Radiocarbon dating of the Church of St. Margaret of Antioch in Kopčany (Slovakia): International consortium results
Switzerland, Zurich, 11.-16.09.2022, 24th Radiocarbon - 10th ^{14}C & Archaeology International Conferences

T. Prášek, M. Němec, M. Christl, P. Gautschi, C. Vockenhuber, M. Kern, P. Steier
Application of fluoride target materials in AMS measurement of uranium isotopes
Czech Republic, Mariánské Lázně, 15.-20.05.2022, 19th Radiochemical conference (RadChem)

U. Ramsperger, L. Wacker, D. De Maria, P. Gautschi, A.M. Müller, S. Maxeiner, H.-A. Synal
LEA – a novel low energy accelerator for ^{14}C dating under a long-term performance test
Switzerland, Zurich, 11.-16.09.2022, 24th Radiocarbon - 10th ^{14}C & Archaeology International Conferences

M. Repasch, J. Scheingross, N. Hovius, R. Szupiany, M. Lupker, N. Haghipour, T.I. Eglinton, D. Sachse
Towards sustainable landscapes-Fluvial organic carbon fluxes modulated by river morphology
USA, Hawaii, 09.-14.07.2022, Goldschmidt 2022

L. Rettig, I. Hajdas, G. Monegato, P. Mozzi, M. Spagnolo
New insights into the last glacial cycle in the south-eastern European Alps from the glacial geomorphological record of the Monte Cavallo (NE Italy)
Austria, Vienna, 23.-27.05.2022, EGU General Assembly 2022

L. Rettig, S. Ivy-Ochs, S. Kamleitner, G. Monegato, P. Mozzi, A. Ribolini, M. Spagnolo
New insights into the Last Glacial Maximum in the Maritime Alps from Equilibrium Line Altitude reconstructions and ^{10}Be surface Exposure Dating
Portugal, Coimbra, 12-16.09.2022, International Conference on Geomorphology 2022

C. Scapozza, D. Giacomazzi, D. Czerski, S. Kamleitner, S. Ivy-Ochs, D. Mazzaglia, N. Patocchi, M. Antognini
Timing of deglaciation and Late Glacial and Holocene infilling of the Ticino valley between Biasca and Lake Maggiore (Southern Switzerland)
Portugal, Coimbra, 12-16.09.2022, International Conference on Geomorphology 2022

M. Schlomberg, C. Vockenhuber, H.-A. Synal
^{32}Si – An alternative radionuclide for dating the recent past?
Switzerland, Zurich, 12.09.2022, 24th Radiocarbon - 10th ^{14}C & Archaeology International Conferences

C. Schnepper, R. Pedrosa-Pàmies, M. Conte, N. Gruber, N. Haghipour, T.I. Eglinton
Tracing bomb radiocarbon in sinking particulate organic carbon in the deep Sargasso Sea
Switzerland, Zurich, 11.-16.09.2022, 24th Radiocarbon - 10th ^{14}C & Archaeology International Conferences

M. Scott, A. Lindroos, G. Barrett, M. Boudin, I. Hajdas, J. Olsen, F. Maspero, F. Marzaioli, D. Michaska, C. Moreau, A. Sironic, F. Pawelczyk
Results and findings from an international mortar dating intercomparison MODIS2
Switzerland, Zurich, 11.-16.09.2022, 24th Radiocarbon - 10th ^{14}C & Archaeology International Conferences

R. Smittenberg, V. Galy, S. Bernasconi, M. Gierga, A. Birkholz, I. Hajdas, L. Wacker, N. Haghipour, C. Ponton, T.I. Eglinton
Terrestrial carbon dynamics through time - insights from downcore radiocarbon dating
Austria, Vienna, 23.-27.05.2022, EGU General Assembly 2022

R. Smittenberg, V. Schwab, M. Gierga, S. Bernasconi, I. Hajdas, L. Wacker, S. Trumbore, X. Xu
Insight in high alpine soil carbon dynamics from compound-specific and soil fraction radiocarbon analysis on a glacier forefield chronosequence
Switzerland, Zurich, 11.-16.09.2022, 24th Radiocarbon - 10th ^{14}C & Archaeology International Conferences

T.J. Suhrhoff, J. Rickli, M. Christl, A. Prokopenko, D. Vance
Reconstructing past weathering conditions at Lake Baikal using radiogenic Sr, Nd, and Pb and meteoric Be isotopes
USA, Hawaii, 09.-14.07.2022, Goldschmidt 2022

H.-A. Synal
AMS and future advances in ^{14}C measurement
United Kingdom, Silverstone, 16.05.2022, Royal Society Meeting

H.-A. Synal
(R) evolutionary developments towards high pecision AMS radiocarbon
Switzerland, Zurich, 11.-16.09.2022, 24th Radiocarbon - 10th ^{14}C & Archaeology International Conferences

H.-A. Synal
Latest developments in Accelerator Mass Spectrometry
Poland, Gliwice, 08.12.2022, Grand opening of the CEMIZ Isotope Methode Center

M. Veicht, I. Mihalcea, P. Sprung, Y. Nedjadi, T. Ramiro, C. Bailat, K. Kossert, D. Symochko, O. Naehle, M. Schlomberg, C. Vockenhuber, S. Röllin, A. Wallner, D. Cvjetinovic, D. Schumann
Towards implementing new isotopes for environmental research: The half-life of ^{32}Si
USA, Sacramento, 21.-29.07.2022, ND2022 – 15th International Conference on Nuclear Data for Science & Technology

C. Vockenhuber
^{32}Si - towards a new half-life measurement
Austria, Vienna, 22.03.2022, VERA Seminar

L. Wacker
Where we don't see any events in ^{14}C
Switzerland, Bern, 07.-10.06.2022, Solar Extreme Events: Setting Up a Paradigm

L. Wacker, N. Brehm, M. Christl, H.-A. Synal, A. Bayliss, K. Nicolussi, C. Pearson, N. Bleicher, D. Brown, S. Bollhalder, M. Alter
Towards a continuous, annually resolved tree-ring record spanning the past 6000 years
Switzerland, Zurich, 11.-16.09.2022, 24th Radiocarbon - 10th ^{14}C & Archaeology International Conferences

A.-M. Wefing, N. Casacuberta, M. Christl
Tracing Arctic Ocean circulation with ^{129}I and ^{236}U
Switzerland, Zurich, 02.12.2022, IBP Seminar (D-USYS)

A.-M. Wefing, N. Casacuberta, M. Christl, P.A. Dodd
Circulation timescales and water mass composition in the Fram Strait determined from the combination of ^{129}I and ^{236}U
Online, 28.02.-04.03.2022, 2022 Ocean Sciences Meeting

A.-M. Wefing, N. Casacuberta, A. Payne
Tracing Arctic Ocean circulation with ^{129}I and ^{236}U
Switzerland, Bern, 02.09.2022, Swiss Polar Day

A.-M. Wefing, A. Payne, N. Casacuberta, M. Christl, J.N. Smith
Circulation timescales and new data on ^{129}I and ^{236}U across the Arctic Ocean
Iceland, Hafnarfjörður, 09.-11.05.2022, Arctic and Sub-arctic Ocean Fluxes Meeting

M. Wertnik, L. Wacker, M. Christl, H.-A. Synal, C. Welte
Data Reduction for Rapid, Continuous Radiocarbon Measurements by Laser Ablation
Switzerland, Zurich, 11.-16.09.2022, 24th Radiocarbon - 10th 14C & Archaeology International Conferences

M. Wertnik, L. Wacker, N. Haghipour, S. Bernasconi, H.-A. Synal, C. Welte
Simultaneous ^{14}C and ^{13}C measurements for any source of CO_2
Switzerland, Zurich, 11.-16.09.2022, 24th Radiocarbon - 10th 14C & Archaeology International Conferences

M. Wertnik, C. Welte, L. Endres, H.-A. Synal, T.I. Eglinton
Recent Improvements in Rapid, Continuous Radiocarbon Measurements by Laser Ablation
Austria, Innsbruck, 17.-20.07.2022, Karst record IX

C. Yeman, M. Christl, R. Witbaard, L. Wacker, B. Hattendorf, C. Welte, H.-A. Synal
Tracking the ^{14}C bomb peak recorded in Arctica Islandica across the North Sea and Northeast Atlantic Ocean
Switzerland, Zurich, 11.-16.09.2022, 24th Radiocarbon - 10th 14C & Archaeology International Conferences

R. Yokochi, R.J. Purtschert, N.C. Sturchio, P. Mueller, S. Wheatcraft, J.A. Corcho Alvarado, N. Duran, M. Leuenberger, S.L. Musy, S. Rollin, C. Vockenhuber
Revisiting the Milk River Aquifer with novel tracers
USA, Hawaii, 09.-14.07.2022, Goldschmidt 2022

A. Zaki, G. King, N. Haghipour, R. Giegengack, S. Watkins, S. Gupta, M. Schuster, H. Khairy, S. Ahmed, M. El-Wakil, S. Eltayeb, F. Herman, S. Castelltort
Intense precipitation during the African Humid Period inferred from east Saharan fossil rivers: Implications for human dispersal
Austria, Vienna, 23.-27.05.2022, EGU General Assembly 2022

G. Zazzeri, P. Gautschi, H. Graven, L. Wacker
A sampling system for ^{14}C analysis of atmospheric methane: from a laboratory prototype to an automated system
Switzerland, Zurich, 11.-16.09.2022, 24th Radiocarbon - 10th 14C & Archaeology International Conferences

M. Zheng, F. Adolphi, C. Paleari, T. Erhardt M. Hörhold, M. Wu, M. Christl, P.Chen, Z. Lu, R. Muscheler
Solar and atmospheric signals in ^{10}Be depositions in Greenland and Antarctica over the last 100 years
Switzerland, Crans-Montana, 02-07.10.2022, International Partnership in Ice Core Sciences - 3rd Open Science Conference

SEMINAR
'CURRENT TOPICS IN ACCELERATOR MASS SPECTRO-METRY AND RELATED APPLICATIONS'

Spring semester

23.02.2022
Yao Gu (ETHZ), Radiocarbon dating of small snail shells at Mangshan section, south-east Chinese Loess Plateau

02.03.2022
Christian Heusser (ETHZ), Improving Compound Specific ^{14}C Analysis for Studies on the Carbon Cycle

09.03.2022
Richard Ott (GFZ, Germany), Decoupling between fluvial aggradation-incision dynamics and paleo-denudation rates during the last glacial cycle, Crete, Greece

16.03.2022
Michael Krzemnicki (SSEF Swiss Gemmological Institute), The Black Prince Ruby and the Pearl of Marie Antoinette: Gem testing in a gemmological laboratory with a special focus on radiometric dating of pearls and gemstones

23.03.2022
Margot White (ETHZ), Ramped pyrolysis/oxidation provides a new perspective on the radiocarbon distribution of marine DOC

30.03.2022
Katrina Gelwick (ETHZ), Quantifying landscape transience using cosmogenic ^{10}Be on the Southeast Tibetan Plateau

06.04.2022
Anne Sofie Søndergaard (ETHZ), Developing the first in-situ C-14 glacial chronology for North Greenland

13.04.2022
Nicolas Brehm (ETHZ), Solar events recorded in trees

27.04.2022
Sascha Maxeiner (Ionplus AG), Low energy accelerator (LEA) - design concepts and measurement performance

04.05.2022
Elena Bruni (ETHZ), Radiocarbon dating fine-grained sediment in oxygen minimum zones

11.05.2022
Melina Wertnik (ETHZ), Universal Interface for simultaneous ^{13}C and ^{14}C measurements

18.05.2022
Giulia Zazzeri (ETHZ/Imperial College London, UK), Quantification of fossil methane emissions using radiocarbon measurements

25.05.2022
Mikko Kivekäs (Univ. Jyväskylä, Finland), Low energy heavy ion elastic scattering and recoiling cross-sections measured by ToF-ERDA

01.06.2022
Elena Chamizo (CNA Seville, Spain), Using the $^{233}U/^{236}U$ atom ratio to trace natural and anthropogenic U in the Arctic Ocean

Fall semester

26.10.2022
Katrina Kremer (SED Swiss Seismological Service), From lake sediment towards a prehistorical Swiss geo-event database for paleoseismology

02.11.2022
Andreas Wiederin (Univ. Vienna, Austria), Isobar separation in the actinide range with ILIAMS

09.11.2022
Patrick Fonti (WSL), A journey into tree-ring formation and its significance for understanding tree-ring proxies

16.11.2022
Kai Nakajima (ETHZ), Application of flow cytometry for radiocarbon analysis of pollen

23.11.2022
Annabel Payne (ETHZ), Everything but the kitchen sink: Anthropo-, Cosmo- and Lithogenic radionuclides to trace water mass transport, mixing and ventilation in the Canada Basin

30.11.2022
Daniele de Maria / Marco Bolandini (ETHZ), Perspective and future applications of CO_2 gas measurements with the double trap interface

07.12.2022
Matthias Schlomberg (ETHZ), Isobar separation of Si-32 from S-32

14.12.2022
Rob Spencer (Univ. of Florida, USA), Radiocarbon Insights into the Changing Cryosphere

21.12.2022
Dominik Fleitmann (Univ. of Basel), Speleothems at the Interface of Climatology, Ecology and Archaeology

THESES (INTERNAL)

Diploma/Master theses

Valentin Gasser
Holocene fluctuations of the Silvretta Glacier (GR, Switzerland) constrained with geomorphological field mapping and cosmogenic nuclide dating
ETH Zurich

Vicente Melo Velasco
Reconstructing the Lateglacial and Holocene glacial history at the Leg Grevasalvas catchment (Graubünden, Switzerland)
ETH Zurich

Joel Scheuchzer
Deciphering the landforms at Cinuos-chel (Graubünden, Switzerland) with geomorphological and sedimentological survey and cosmogenic ^{10}Be dating of boulders
ETH Zurich

Doctoral theses

Sarah Maria Kamleitner
Reconstructing the evolution and dynamics of Central Alpine glaciers during the Last Glacial Maximum on the basis of their geomorphological footprints and cosmogenic nuclide surface exposure dating
ETH Zurich

THESES (EXTERNAL)

Doctoral theses

Joshua Chaplin
Development and application of novel DGT sampling technologies to monitor bioavailable actinide species
University of Lausanne

Silvia Pérez-Diez
Red cinnabar blackening in the mural paintings of Pompeii: volcanic impact, atmospheric exposure and burial environment
University of the Basque Country UPV/EHU, Spain

Jesper Surhoff
Modern and past chemical weathering at Lake Baikal and its implications for marine reconstructions of the global weatheringclimate feedback
ETH Zurich

Mario Aaron Veicht
Towards Implementing New Isotopes for Environmental Research: Redetermination of the ^{32}Si Half-Life
EPFL Lausanne

External Referee

Lin Mu
Anthropogenic U-236 and U-233 in the Baltic Sea: Sources, Distributions, and Tracer Application
Technical University of Denmark, Denmark

COLLABORATIONS

Australia

The Australian National University, Research School of Physics, Canberra

University of New South Wales, Earth Science, Sydney

Austria

AlpS - Zentrum für Naturgefahren- und Riskomanagement GmbH, Geology and Mass Movements, Innsbruck

Austrian Academy of Sciences, Institute for Oriental and European Archaeology, Vienna

Geological Survey of Austria, Sediment Geology, Vienna

International Atomic Energy Agency, Vienna

University of Innsbruck, Institute of Geography, Geology and Botany, Innsbruck

University of Salzburg, Geography and Geology, Salzburg

University of Vienna, VERA, Faculty of Physics, Vienna

Vienna University of Technology, Institute for Geology, Vienna

Belgium

Royal Institute for Cultural Heritage, Brussels

Université catholique de Louvain, Earth and Life Institute, Louvain-la-Neuve

Bermuda

Bermuda Institute of Ocean Sciences, Bermuda

Canada

Bedford Institute of Oceanography, Bedford Institute of Oceanography, Halifax

Chalk River Laboratories, Dosimetry Services, Chalk River

China

China Institute for Radiation Protection, Dosimetry Services, Taiyuan City

Croatia

IZOR, Institute of Oceanography and Fisheries, Split

Czech Republic

Czeck Technical University, Nuclear Chemistry, Prague

Denmark

Danfysik A/S, Taastrup

Risø DTU, Risø National Laboratory for Sustainable Energy, Roskilde

University Southern Denmark, Department of Physics, Chemistry and Pharmacy, Odense

Finnland

University of Jyväskylä, Accelerator Laboratory, Physics Department, Jyväskylä

France

Aix-Marseille University, Collège de France, Aix-en-Provence

Commissariat à l'énergie atomique et aux énergies alternatives, Laboratoire des Sciences du Climat et de l'Environnement (LSCE), Gif-sur-Yvette Cedex

Université de Savoie, Laboratoire EDYTEM, Le Bourget du Lac

Germany

Alfred Wegener Institute of Polar and Marine Research, Marine Geology, Bremerhaven

Berlin-Brandenburgische Akademie der Wissenschaften, Berlin

Bonn University, Steinmann Institute, Bonn

GFZ German Research Centre for Geosciences, Klimadynamik, Dendrochronology Laboratory, Earth Surface Geochemistry, Potsdam

Helmholtz-Zentrum Dresden-Rossendorf (HZDR), Institute of Ion Beam Physics and Particle Physics, Dresden

Hydroisotop GmbH, Schweitenkirchen

IFM-GEOMAR, Palaeo-Oceanography, Kiel

Leibniz Institute for the History and Culture of Eastern Europe (GWZO), Leipzig

Leibniz-Institut für Ostseeforschung , Marine Geology, Warnemünde

Physikalisch Technische Bundesanstalt, Fachbereich Radioaktivität, Braunschweig

Potsdam-Institut für Klimafolgenforschung, Complexity Science, Potsdam

Reiss-Engelhorn-Museen, Curt-Engelhorn-Zentrum Archäometrie gGmbH, Mannheim

RWTH Aachen, Engineering Geology, Aachen

University of Applied Sciences, TH Köln, Technology Arts Sciences, Cologne

University of Cologne, Institute of Geology and Mineralogy, Cologne

University of Hannover, Institute for Radiation Protection and Radioecology, Hannover

University of Heidelberg, Institute of Environmental Physics, Geosciences, Heidelberg

University of Hohenheim, Institute of Botany, Stuttgart

University of Tübingen, Department of Geoscience, Early Prehistory and Quaternary Ecology, Archaeozoologie, Tübingen

Hungary

Hungarian Academy of Science, Institute of Nuclear Research (ATOMKI), Debrecen

India

Indian Institute of Science Education & Research Kolkata Department of Earth Sciences, Nadia, West Bengal

Inter-University Accelerator Center, Accelerator Division, New Dehli

Italy

Geological Survey of the Provincia Autonoma di Trento, Landslide Monitoring, Trento

INGV Istituto Nazionale di Geofisica e Vulcanologia, Sez. Sismologia e Tettonofisica, Rome

Südtiroler Landesverwaltung, Land- und Forstwirtschaft, Bolzano

University of Bologna, Department of Biological, Geological and Environmental Sciences (BiGeA), Bologna

University of Padua, Department of Geosciences, Geology and Geophysics, Padua

University of Pisa, Department of Earth Science, Pisa

University of Salento, Department of Physics, Lecce

University of Turin, Department of Geology, Turin

Japan

University of Tokai, Department of Marine Biology, Tokai

Liechtenstein

Oerlikon Surface Solutions AG, Balzers

Monaco

International Atomic Energy Agency, IAEA Environment Laboratories

New Zealand

University of Waikato, Radiocarbon Dating Laboratory, Waikato

Victoria University of Wellington, School of Geography, Environment and Earth Sciences, Wellington

Norway

Norwegian Geological Survey, Trondheim

Norwegian Polar Institute, Tromso

Norwegian Radiation Protection Authority, Tromso

Norwegian University of Science and Technology, Physical Geography, Trondheim

Rogaland Fylkeskommune, Stavanger

The Bjerkness Centre for Climate Res., Bergen

The University Museum of Bergen, Bergen

Univeristy of Oslo, Department of Archaeology, Conservation and History, Oslo

Poland

Adam Mickiewicz University, Department of Geology, Climate Change Ecology Research Unit, Poznan

AGH University of Science and Technology, Department of Geology, Kraków

Polish Academy of Sciences, Institute of Geography and Spatial Organization, Warsaw

University of Marie Curie Sklodowska, Department of Geography, Lublin

Romania

Horia Hulubei - National Institute for Physics and Nuclear Engineering, Magurele

Singapore

National University of Singapore, Department of Chemistry, Singapore

Slovakia

Comenius University, Faculty of Mathematics, Physics and Infomatics, Bratislava

Spain

Consejo de Investigaciones Cientificas, Instituto de Investigaciones Marinas de Vigo, Vigo

University of Murcia, Department of Plant Biology, Murcia

University of Seville, Physics Department and National Center for Accelerators, Seville

University of the Basque Country (UPV/EHU), Department of Analytical Chemistry, Vitoria-Gasteiz

Sweden

Lund University, Geology, Lund

Switzerland

Amt für Kultur Kanton Graubünden, Archäologischer Dienst, Chur

Centre Hospitalier Universitaire Vaudois, Institut de radiophysique, Lausanne

Dendrolabor Wallis, Brig

EAWAG, Dübendorf

Empa, Energy Conversion, Advanced Materials Processing, Nanoscale Materials Science, Mechanics of Materials and Nanostructures, Corrosion and Joining Technology, Thin Films and Photovoltaics, Dübendorf

ENSI, Brugg

ETH Zurich, Laboratory for Multifunctional Materials, Institute of Geology, Department of Environmental System Sciences, Environmental Physics, Laboratory of Inorganic Chemistry, Building Research and Construction History, Laboratory of Inorganic Chemistry, Inorganic Chemistry, Engineering Geology, Geological Institute, Institute of Geochemistry and Petrology, Atmospheric Chemistry, Metals Research, Department of Earth Sciences, Earth Sciences, Zurich

Evatec AG, Trübbach

Geneva Fine Art Analysis Sarl, Lancy, Geneva

Haute école d'ingénierie et d'architecture de Fribourg, Fribourg

Kanton Graubünden, Archäologischer Dienst, Chur

Kanton St. Gallen, Kantonsarchäologie, St. Gallen

Kanton Turgau, Kantonsarchäologie, Frauenfeld

Kanton Zug, Kantonsarchäologie, Zug

Kanton Zürich, Kantonsarchäologie, Dübendorf

Labor für quartäre Hölzer, Affoltern a. Albis

Laboratiore Romand de Dendrochronologie, Cudrefin

Nationale Genossenschaft für die Lagerung radioaktiver Abfälle, Wettingen

Office et Musée d'Archéologie Neuchatel, Neuchatel

Paul Scherrer Institut (PSI), Materials Group, Labor für Radio- und Umweltchemie, Mesoscopic Systems, Villigen

Stadt Zürich, Amt für Städtebau, Zurich

Stiftung Pro Kloster St.-Johann, UNESCO Welterbe, Müstair

SUPSI, Dipartimento ambiente costruzioni e design (DACD), Lugano

Swiss Federal Institute for Forest, Snow and Landscape Reseach (WSL), Landscape Dynamics, Dendroecology, Soil Sciences, Birmensdorf

Swiss Gemmological Institute – SSEF, Basel

Swiss Institute for Art Research SIK ISEA, Zurich

University of Basel, IPAS, IPNA, Departement Altertumswissenschaften, Basel

University of Bern, Institute of Geology, Climate and Environmental Physics, LARA, Oeschger Center for Climate Research, Bern

University of Freiburg, Faculty of Environmentat and Natural Resources, Freiburg

University of Geneva, Department of Anthropology and Ecology, Geneva

University of Lausanne, Department of Geology, Lausanne

University of Zurich, Abteilung Ur- und Frühgeschichte, Institute of Geography, Institut für Evolutionäre Medizin, Zurich

Taiwan

National Taiwan University, Department of Geosciences, Taipei

The Netherlands

Twente University, Nanotechnology, Twente

Türkiye

Dokuz Eylül University, Department of Geological Engineering, Izmir

Istanbul Technical University, Faculty of Mines, Istanbul

Tunceli Üniversitesi, Geology Department, Tunceli

United Kingdom

Brithish Arctic Survey, Cambridge

Cambridge University, Department of Earth Sciences, Geography, Cambridge

Durham University, Department of Geography, Durham

Histrorical England, London

Imperial College London, Faculty of Natural Sciences, Department of Physics, London

Queen Mary University of London, School of Geography, London

University of Aberdeen, School of Geosciences, Aberdeen

University of Bristol, School of Chemistry, Bristol

University of Lancaster, Nuclear Engineering, Lancaster

University of Leicester, School of Geography, Leicester

USA

NOAA Fischeries, Pacific Islands Fisheries Science Center, Honolulu

Stanford University, Jasper Ridge Biological Preserve, Woodside, California

University of Utah, Geology and Geophysics, Salt Lake City

Woods Hole Oceanographic Institution, Center for Marine and Environmental Radioactivity, National Ocean Sciences Accelerator Mass Spectrometry, Woods Hole

VISITORS AT THE LABORATORY

Yao Gu
Nanjing University, Nanjing, China
01.01.2022 - 19.07.2022

Fatima Pawalczyk
Silesian University of Technology, Gliwice, Poland
17.01.2022 - 31.03.2022

Kateřina Fenclová
Czech Technical University, Prague, Czech Republic
31.01.2022 - 08.02.2022

Tomas Prasek
Czech Technical University, Prague, Czech Republic
31.01.2022 - 08.02.2022

Stanislav Kolesov
Academy of Science of the Republic of Sakha, Yakutsk, Russia
21.03.2022 - 21.04.2022

Kateřina Pachnerová Brabcová
Nuclear Physics Institute of the CAS, Prague, Czech Republic
28.03.2022 - 08.04.2022

Mikko Kivekäs
University of Jyväskylä, Jyväskylä, Finland
01.04.2022 - 30.06.2022

Jan Valek
The Czech Academy of Sciences, Institute of Theoretical and Applied Mechanics, Prague, Czech Republic
26.04.2022 - 29.04.2022

Petr Kozlovcev
The Czech Academy of Sciences, Institute of Theoretical and Applied Mechanics, Prague, Czech Republic
26.04.2022 - 29.04.2022

Kristýna Kotková
The Czech Academy of Sciences, Institute of Theoretical and Applied Mechanics, Prague, Czech Republic
26.04.2022 - 29.04.2022

Sandro Rossato
University of Padua, Padua, Italy
16.05.2022 - 28.05.2022

Elena Chamizo
Universidad de Sevilla, Seville, Spain
30.05.2022 - 03.06.2022

Lukas Rettig
University of Padua, Padua, Italy
22.06.2022 - 22.07.2022

Tomas Prasek
Czech Technical University, Prague, Czech Republic
19.09.2022 - 30.11.2022

Beatrice Behlen
Museum of London, London, United Kingdom
28.09.2022 - 30.09.2022

Jane Malcolm-Davies
Globe Institute & Centre for Textile Research&The Tudor Tailor/JMD&Co, Copenhagen, Denmark
28.09.2022 - 30.09.2022